Extremes _{of} Mobility

Development and
Consequences of Transport
Policy in Los Angeles

Extremes
of Mobility
Development and Consequences of Transport Policy in Los Angeles

Stefan Bratzel

University of Applied Sciences in Bergisch Gladbach, Germany

 Springer

 World Scientific

Published by

World Scientific Publishing Co. Pte. Ltd.

5 Toh Tuck Link, Singapore 596224

USA office: 27 Warren Street, Suite 401-402, Hackensack, NJ 07601

UK office: 57 Shelton Street, Covent Garden, London WC2H 9HE

Library of Congress Cataloging-in-Publication Data
Names: Bratzel, Stefan, author.
Title: Extremes of mobility : development and consequences of transport policy in Los Angeles /
    Stefan Bratzel, University of Applied Sciences in Bergisch Gladbach, Germany.
Other titles: Extreme der Mobilität. English
Description: New Jersey : World scientific, [2024] | Includes bibliographical references and index.
Identifiers: LCCN 2023027044 | ISBN 9789811278433 (hardcover) | ISBN 9789811278440 (ebook) |
    ISBN 9789811278457 (ebook other)
Subjects: LCSH: Urban transportation--California--Los Angeles. | Local transit--California--
    Los Angeles. | Transportation and state--California--Los Angeles.
Classification: LCC HE309.C2 B7313 2024 | DDC 388.409794--dc23/eng/20230608
LC record available at https://lccn.loc.gov/2023027044

British Library Cataloguing-in-Publication Data
A catalogue record for this book is available from the British Library.

First published in German under the title
Extreme der Mobilität; Entwicklung und Folgen der Verkehrspolitik in Los Angeles
by Stefan Bratzel, edition: 1
Copyright © Springer Basel AG, 1995*
This edition has been translated and published under licence from
Springer Fachmedien Wiesbaden GmbH, part of Springer Nature.
Springer Fachmedien Wiesbaden GmbH, part of Springer Nature takes no
responsibility and shall not be made liable for the accuracy of the translation.

For any available supplementary material, please visit
https://www.worldscientific.com/worldscibooks/10.1142/13474#t=suppl

Desk Editors: Aanand Jayaraman/Yulin Jiang

Typeset by Stallion Press
Email: enquiries@stallionpress.com

Printed in Singapore

For Suzanne, Friso, and Sophie

who endure me patiently on my drive through
Los Angeles and through life

Preface (First Edition)

This book is about mobility experiences in Los Angeles in a double sense. According to my experience in Los Angeles, the bicycle and feet do not provide a complete picture of mobility there. In order to really understand the different dimensions of mobility and its importance for the citizens, one has to experience the metropolis — with an automobile. The attentive ride on one of the numerous multi-lane freeways is not only a kinetic experience whose joys and sorrows can be shared with thousands of people in Los Angeles at any time of day. The freeway experience is much more: Joan Didion rightly describes it as "the only secular creed that Los Angeles has" (Didion 1977). The freeways are the metropolis's *automotive churches*, the city's landmarks, and symbols of mobility. The sociocultural orientations and lifestyles of the citizens of southern California are condensed in them, but increasingly also the problems of modern urban agglomerations. More and more, accidents, congestion, and high environmental pollution are shaking the citizens' *belief* in (auto)mobility.

Los Angeles is an extreme case in terms of transport policy, from which the negative consequences of mobility can be read in an ideal way. Moreover, the history of the metropolis's development offers the opportunity to more clearly identify the causes and interactions of mobility and transport policy, precisely because, compared to other metropolises, it has not been fundamentally different, but only more extreme. Other cities can learn from the patterns with which Los Angeles reacted and continues to react to its considerable transport policy problems, especially in order to avoid the mobility experiences of Los Angeles. However, to do so, it is necessary to abandon the prevailing perspective, which is largely limited to evaluating transport modes ("road versus rail"). A sustainable solution to the problems of metropolitan areas will only be possible if the origins of transport with their spatial, social, and economic interdependencies are understood and integrated into (transport) policy action.

The book is the result of a research trip to California and an earlier version was accepted by the Free University of Berlin as a diploma thesis. In this context, I would like to thank all those who supported the realization of this work. For helpful suggestions and information on site, I would like to thank in particular Joel Woodhull (Southern California Rapid Transit District), Martin Wachs (University of California Los Angeles), Ralph Cipriani (Southern California Rapid Transit District), Cheryl Collier and Dominic Menton (Commuter Transportation Services), as well as Stanley Hart (Sierra Club). The work was challenged by a variety of theoretical and conceptual references, for which I owe a debt of gratitude above all to Volker von Prittwitz (Free University of Berlin) and Martin Jänicke (Free University of Berlin). For helpful criticism and suggestions regarding content, I would like to thank Kai Wegrich (Humboldt University Berlin) in particular. I would also like to thank the Berlin Senate Chancellery, Birkhäuser Verlag, and Helmut Wollmann (Humboldt University Berlin) for their unbureaucratic support of the research project. For their help in the technical realization of the book, I would like to thank Ani Kroeger, Barbara Klebe, and Sassa Franke. Finally, I would like to thank my colleagues from the Environmental Policy Research Centre at the Free University of Berlin for the fruitful exchange of ideas and the pleasant working atmosphere. Of course, I alone bear the responsibility for any shortcomings in content or form.

Berlin, December 1994

Preface (New/Expanded Edition)

In the first edition of this book, I tried to describe the history of the emergence of motorized mobility and traffic in Los Angeles between 1890 and 1990. In a 100-year longitudinal comparison, influencing factors and reasons that have made Los Angeles an extreme case of a car-centric metropolis were discussed. In this new edition, the study of the transport policy of the metropolis over the last 30 years has been added, in which the (re)construction of a large public rail transport network in Los Angeles was the political focus. The questions I was interested in included the following: Has politics introduced a promising change of strategy in transport policy that really solves the traffic problems in the metropolis? Is the public transit system going to be used and accepted by the population of Los Angeles, and does this result in significant changes of mobility behavior? What effects will the age of autonomous driving have on mobility in Los Angeles?

Los Angeles is a fascinating metropolis and closely linked to my professional career. As a young student in Los Angeles in the early 1990s, I really wanted to make my childhood dream come true and do a few pull-ups on the Hollywood "H". However, as I gazed from the Hollywood "H" at the smog and congested freeways of L.A., a deep interest in the connections between mobility and transport policy arose. After completing the study on the "car metropolis Los Angeles", I examined — as a contrast program — "success cases" of environmentally oriented transport policies in European cities such as Amsterdam, Zurich, and Freiburg, which have achieved a relatively low share of cars in total traffic. It was shown that an ecological change in transport policy in all cities was only possible through a presuppositional and mostly very conflict-ridden political process (Bratzel 1999a, 1999b). Over the years, I have visited California and Los Angeles for work and pleasure regularly. And then, when I was stuck in traffic jams on the Los Angeles freeways and looked at the prominent Hollywood sign, I realized, "Progress is a snail".

In this new edition of the book, I have added a completely new chapter ("Back to the Future: L.A. Revisited 1990–2022"), in which the transport policy of the last 30 years is illuminated against the background of the theses of the book. The previous Chapters 1–6 of the first edition, which examine the development of transport and mobility between 1890 and 1990, have only been translated into English and have otherwise not been modified.

I would like to thank the publisher World Scientific Publishing, who encouraged me to reprint the book. Special thanks go to Dr. Annalisa Fischer, who did not give up although I was unable to meet the agreed deadlines. Further, I would like to thank Felix Boebber who supported me in the technical realization of the book. And of course a big hug to my wife and children, who accompanied me on my journeys through Los Angeles and through life.

Bergisch Gladbach/Cologne, February 2023

About the Author

Stefan Bratzel (born 1967) studied Political Science at the Freie Universität Berlin. His Ph.D. thesis "Success conditions of environmentally oriented transport policy" was recognized by the Foundation of German Cities, Municipalities, and Districts. Bratzel worked in and around the auto industry. As a Product Manager in the Daimler-Chrysler Group, he helped shape the beginnings of connected cars and mobility services. In 2004, as a Professor of Automotive Management, he became Founder and Director of the Research Institute Center of Automotive Management (CAM) that focuses on studies of innovations in the automotive industry and future trends in (auto)mobility.

Contents

List of Figures

List of Tables

List of Abbreviations

AVR   Average Vehicle Ridership
CBD   Central Business District
CMSA   Consolidated Metropolitan Statistical Area
GM   General Motors
L.A.   Los Angeles
LACTC   Los Angeles County Transportation Commission
LAMTA   Los Angeles Metropolitan Transit Authority
LARY   Los Angeles Railway
METRO   Los Angeles County Metropolitan Transportation Authority
NCL   National City Lines
PE   Pacific Electric Railroad
PMSA   Primary Metropolitan Statistical Area
SCAG   Southern California Association of Governments
SCAQMD   South Coast Air Quality Management District
SCRTD   Southern California Rapid Transit District
SMSA   Standard Metropolitan Statistical Area
TRIP   Transportation Reduction Improvement Program
TSM   Traffic System Management

Chapter 1

Introduction

To a man with a hammer, all problems look like nails.

Old Japanese proverb

1.1 Mobility and transport — The thesis

The background to this work is the observation that in the debates on *urban mobility,* simple answers to complex questions are often given prematurely, which makes it considerably more difficult to deal with transport policy problems in a sustainable manner. Instead of first grasping the respective problem in all its complexity, then looking at the fundamentally possible political options for action and finally developing effective strategies for action, an attempt is usually made to apply old solutions to dynamically changing new problems.

Accordingly, discussions of the topic "city and traffic" not only in public but also in science are characterized by a tension-filled dichotomous juxtaposition of the supposed antagonists — "automobile" and "railway". The *cause of* urban transport problems is now generally seen to be car traffic, while the (yet to be realized) *solution to* the problems is represented by the railway or local public transport. The "environmental friendliness of the railways" (Weizäcker 1990: 86) and their objective performance is often contrasted with the inefficiency and delusional character of car and road transport.[1] All sections of the population, it is

---

[1] Wilfried Wolf's "Eisenbahn und Autowahn" (1987) expresses not only the dichotomy of modes of transport but also the irrational behavior attributed to car traffic (alone).

stressed, have already learnt the argument of the environmental friendliness of public transport (Müßener 1991: 447).

The aim of this work is to counter the widespread opinion that the "rail mode of transport" is per se the solution to the mobility problems of urban agglomerations. Rather, this work will show that the modes of transport — car/road and train/rail — can in *principle* have *similar effects* on the mobility structures (spatial arrangement of the activities living, work, consumption, and leisure) and the mobility behavior of citizens in a metropolis. Distance-intensive mobility structures and the corresponding mobility behavior can be brought about or promoted not only by investments in road and motorway construction but also by extensive expansion of rail networks.

If these dominant supply-oriented political patterns of action are not changed, but instead applied to the *new* transport mode "rail" instead of road construction, there is a great danger that it will not be possible to guarantee environmentally compatible mobility in metropolitan areas. A sustainable solution to the mobility problems of cities can only be expected if the "mode of transport orientation" in transport policy is abandoned and the conditions under which transport arises, with its social and cultural implications, are incorporated into political action.

1.2 Los Angeles: Extremes of mobility

The negative consequences of a predominantly supply-oriented transport policy can be seen in the development of the metropolis of Los Angeles, which is in many respects an extreme case. The city's extreme development patterns nevertheless offer the opportunity to more clearly identify the interrelationships and interactions between mobility, sociocultural conditions, and political action.

Although the transport policy in Los Angeles has not been fundamentally different from that of other cities, the key figures of its development history and its consequences are unmistakable in their conciseness:

- At the beginning of the 20th century, Los Angeles had the largest electric railway network in the world.

Similarly, Wolfgang Sachs 1984, "Die Liebe zum Automobil" or Gerhard Armanski 1986, "Die Lust und Last am Automobil".

- In the 1920s and for many years afterward, the metropolis had by far the highest car motorization rates in the world, i.e., automobiles per inhabitants.
- With the spread of the automobile, the citizens of the metropolis increasingly turned away from rail transport. By 1963, all tram and regional railway lines had been closed down.
- Since the end of the 1980s, "the most motorized region in the world" (Bacharach 1991: 7) has been investing $78 billion(!) in an unprecedented investment program to rebuild a huge rail-bound public transport system.

Another extreme characteristic of L.A. is the *low-density* lifestyle of its citizens, the *Angelenos*, which has shown remarkable consistency over the various phases of transport policy development. These behavioral dispositions have been structurally reflected in the extremely low settlement density compared to the other metropolises of the USA. The more than 14.5 million inhabitants of the metropolis Los Angeles (Consolidated Metropolitan Statistical Area (CMSA) Los Angeles–Anaheim–Riverside) are spread over a huge area of 88,000 sq. km. The largest metropolis in terms of population, New York (18 million inhabitants), on the other hand, has an area of 18,000 sq. km, only a good fifth of the size of L.A. Even in the densely populated core area of Los Angeles — Long Beach (Primary Metropolitan Statistical Area, PMSA) — where almost 9 million people live, a comparatively modest density of 842 inhabitants/sq. km is achieved (New York PMSA: 2875 sq. km) (cf. Table 1.1).

However, Los Angeles is regarded as an extreme case, especially with regard to the realization of a certain understanding of mobility.

Table 1.1:   **Population, area, and settlement density of the largest metropolitan areas in the US (1990).**

Metropolis (CMSA)	Population (billions)	Area (sq. km)	Density (inhabitants/sq. km)
New York	17.93	18,134	990
Los Angeles	14.53	87,970	165
Chicago	8.07	14,553	554
San Francisco	6.25	19,085	327
Philadelphia	5.9	13,844	426

Source: Hall/Slater (1992).

The metropolis is *the prototype and symbol of an automotive society*. The dependence of citizens on cars and the social and ecological consequences of this automobility are probably more obvious here than in any other metropolis. Whereas in Los Angeles in 1980 over 87% of employed people used the car to get to work, only 61% did so in New York, 68% in Boston, 76% in Chicago, and 79% in San Francisco (Bureau of the Census 1986). The backbone of the region's transport system is the network of freeways, some 2000 miles long, which are rightly called the secular cathedrals of Southern California. A majority of passenger kilometers are traveled on these monuments, which often consist of 14 to 16 lanes and are comparable to urban motorways. However, the Angelenos, with their more than 40 million daily journeys, also face traffic jams for about 10% of their above-average journey time.

The ecological burdens of the auto orientation of Southern California are equally extreme. Despite strict exhaust emission regulations for cars, Los Angeles has by far the worst air quality in the US. Although the situation has improved since the 1960s, Los Angeles still violated state air quality standards for ozone on 130 days in 1990 and California's air quality standards for ozone on 180 days (South Coast Air Quality Management District 1991).

The high level of (auto)mobility of the Angelenos becomes obvious when unexpected disruptions in the transport network occur. For example, the earthquake in January 1994, which caused the collapse of two bridges on the busiest freeway in Southern California (Santa Monica Freeway), made people painfully aware of their distance-intensive mobility behavior and their dependence on cars. Whereas commuters had been able to cover long distances on the freeways in a calculable time, travel times after the earthquake rose from half an hour to 3–5 hours for the easy way to work (*New York Times*, 26.1.1994). For the first time, many workers were forced to give up the car and look for alternatives.

From a transport policy point of view, the disaster had positive effects in this respect. The impassable freeways triggered sudden pressure to act and revived ideas that had already been developed. As a result of the earthquake, the possibilities of teleworking in particular began to be exploited: Employers provided PCs and teleconferencing systems so that their employees could work from their homes. *Satellite workstations* were increasingly set up close to the homes of employees. In addition to providing incentives for more flexible working hours (e.g., a four-day week), employers also demanded *job tickets* for the metropolis's new commuter trains, which were relatively unaffected by the earthquake. After the catastrophe, the morning passengers on the commuter trains tripled from 10,000 to 30,000 and for the first time ensured reasonable utilization of the otherwise hardly used suburban trains (*Time Magazine*, 14.2.1994). Nevertheless, the restoration of the freeways months later also largely restored the old mobility behavior of citizens.

The transport policy history of the extreme case of Los Angeles offers, beyond the troublesome events of these years, fruitful illustrative material for understanding basic processes and interrelationships of mobility. The high level of mobility among citizens is not, for example, the linear consequence of the explosive spread of the automobile in the 1920s and after the Second World War. Rather, the main reasons for this are to be found in the reduction in spatial resistance caused by trams and regional railways, which at the turn of the century created the basic spatial structures of today's metropolis.

In the course of this book, it will be shown how the interaction of transport modes, mobility structures, mobility behavior, and the political action of the actors resulted in distance-intensive transport relations, which ultimately led to the creeping mobility compulsion of citizens. The construction of large-scale transport systems (rail networks or highways) in Los Angeles initially expanded the mobility opportunities of citizens by significantly increasing the spatial accessibility of the region. Accessibility, however, led to systemic adaptation processes, resulting in changes in mobility behavior and structures: At the same time, it was possible to choose distant activity locations, the remote "house in the country" became more attractive, the conditions for setting up businesses "on greenfield sites" (near motorways) were improved, etc. This led to a general increase in the mobility level of the population as a whole, which inadvertently transformed mobility opportunities into mobility needs.

The case of Los Angeles is also a good example of the influence transport policy has or can have on these processes, for example, the devastating effect that political *inaction* can have. A momentous misunderstanding was the assessment of many actors in the 1920s that the chaotic traffic conditions in the center of the metropolis, which were mainly caused by cars, would force citizens to switch back to public transport even without state intervention. However, instead of doing without a car, citizens increasingly avoided the problem area of the center and successively chose other destinations. The trade and service sectors also shifted their locations to other, less polluted areas. In this respect, it is important to warn against the widespread misunderstanding that — first — sufficient pressure from problems automatically leads people to rethink and that — second — a high level of pollution simultaneously increases the capacity to deal with problems. Problem pressure seems at best to be a necessary, but by no means sufficient, condition for both awareness processes and political action.

Finally, the extreme transport policy case of Los Angeles illustrates some aspects of the often neglected spatial–social dimension of mobility. Both railways and automobiles, for example, which are actually intended to bridge spaces and bring people together, often have precisely the opposite, decoupling effect: They are the means of avoiding social contact with undesirables and can cement social differences. Moreover, the opportunities for mobility are distributed very differently between the various social groups. In Los Angeles, it is mainly middle- and upper-income citizens who benefit from the high level of mobility, while the social isolation of ethnic minorities and low-income citizens has become more entrenched.

1.3 The analysis concept: Mobility patterns

The following outlines the analytical concept of "mobility patterns" that underlies this study. First of all, however, the central terms "mobility" and "transport" are briefly explained, which — although they are very often used in trivial and academic literature — are rarely defined in terms of content.

The term transport is usually only used in connection with *real* movement processes, whereas mobility (from Latin: mobilis (movable)), on the other hand, also stands for the *ability* or *possibility* (capacity) of individuals and groups to carry out movements. "Transport" is defined, for example, as "the spatial movement of persons (passenger traffic), goods (freight traffic) and communications (communication) *using special technical and organisational facilities* (means of transport, traffic routes, stations and installations)" (Hervorh., S.B.) (Brockhaus 1974:511). In contrast, the term "mobility" is mostly used only in connection with individuals, social groups, or even societies or their (mobility) behavior.

Jones stresses, "(I)t is evident that we need to understand mobility as relating to travel and non-travel aspects of behaviour, both in order to gain a full understanding of the concept and to be able to set it in the context of the individual's daily life" (1987: 45). In this sense, mobility can be defined, following Cerwenka, as a measure "of the ability or capacity of individuals and groups to perform movements and to overcome distances, in terms of their technical, economic and physical potency" (1982: 8).

When individuals move in space, traffic is created. Four recurring reasons for movement or mobility can be distinguished: living, work, consumption (supply), and leisure. The further apart the places for these activities are from each other, the more space has to be overcome and the more *traffic* is generated (cf. Hägerstrand 1987; Cerwenka 1982; Meyfahrt 1987: 115 ff). The concept of *mobility patterns*, which forms the basis of this study and is explained in more detail in the following, is based on the conditions under which traffic arises. Mobility patterns are a syndrome of three elements at different levels: mobility structure, mobility behavior, and mobility opportunities.

The *mobility structure* is the basic level of mobility. This term refers to the spatial arrangement of the places of activity (living, working, consumption, and leisure). The mobility structure forms the relatively permanent basis for the mobility behavior of individuals and changes only in the medium and long terms. Empirically, the mobility structure can be determined at the predominant distribution of places of residence, work, consumption, and leisure facilities in a region. The population density, type of settlement (housing structures), and the patterns of land use (type, form, and density of use) also form indicators for the mobility structure.

The *mobility behavior* of individuals is based on the mobility structure. Thus, the distance between the places of activity determines the choice of transport mode. While distance-intensive mobility structures make the use of motorized modes of transport unavoidable, dense structures also allow the use of bicycles and feet (cf., e.g., Monheim/ Monheim-Dandorfer 1990: 180–346). Within the framework of the mobility structure and the availability of certain modes of transport (car ownership, public transport connections), the individual has scope for decision-making. They can, for example, choose nearby destinations for their activities (shopping, leisure time, etc.) or (for this purpose) *go to the countryside*.

If many people change their mobility behavior, for example, by choosing a different place of residence, the entire mobility structure may also change in the longer term (Noortman 1978: 9). Mobility behavior can be measured using aggregated data such as the modal split (the modal split refers to the choice of transport mode for the distances traveled).[2] This is done by determining the (average) number of journeys (per time unit), the journeys' length, and the (journey) time required for this (Cerwenka 1982: 10 ff; Jones 1987: 35 ff; Zumkeller 1987; Der Sachverständigenrat 1994).

Very often, it is not taken into account that mobility has a *socio-psychological dimension* that should not be underestimated, beyond the (purpose — rational) possibility of reaching different places of activity. Overcoming space is often an end in itself, demands well-being, and can produce an "optimal level of excitement for the organism" (ADAC 1987). Psychologists also confirm that "beyond all considerations of usefulness and psychological motives (...) movement is a very original basic need" (Nowak 1987: 17). Psychologically, mobility offers the potential for escape, which is often exploited: from performance requirements, as compensation for missed experiences (Ullrich 1988: 26 f), from the complexity of the environment, from the confinement and social pressure of the family or village community — "our means of transport are all vehicles for small escapes" (Nowak 1987: 14).

The third element of mobility patterns is the *mobility opportunities of social groups*. The central question here is which social groups are less mobile and why. Only since the end of the 1960s has there been a more intensive discussion of the fact "that mobility is always a selective process: it discriminates according to age, income, race and many other attributes" (Nijkamp 1987: 80 f).[3]

Mobility opportunities, in turn, vary considerably according to the respective mobility structures and accessibility to modes of transport. For example, mobility structures that are intensive in terms of distance put older people, the disabled, children, and women at a particular

---

[2]Characteristically, as for the auto-oriented development of transport policy after the Second World War, statistical studies of the choice of means of transport in the USA and also in Europe were often limited to motorized means of transport and thus ignored the mobility potential of bicycle and feet (Meyer/Gomez-Ibanez 1981: 22 f; Monheim/ Monheim-Dandorfer 1990: 89–94).

[3]Following the race riots in Los Angeles in the 1960s, the selective dimension of mobility was discussed intensively for the first time in America. At that time, a Commission report pointed out that the structure of the city and the transport system could be factors which segment the poverty of certain sections of the population (McCone 1966; Altshuler 1979: 274; see below).

disadvantage (Paaswell/Recker 1978; Altshuler 1979: 252 ff; Monheim/ Monheim-Dandorfer 1990: 109–112). The elderly and disabled, for example, are considerably more dependent on public transport systems, especially in conurbations (*ibid.:* 111; Altshuler, 1979: 260 ff). The mobility forced by distance-intensive structures disproportionately affects children, women, and old people, who otherwise walk far more than the average of the total population (Monheim/Monheim-Dandorfer 1990: 112).

A sensitive comparison of *general* mobility behavior data with those of specific *social groups* is a good indicator of mobility opportunities. Often — as it is sometimes the case in Los Angeles — such specific data are not available. Particularly in the case of mobility structures that are intensive in terms of distance and poor local public transport systems, the availability of cars can also be used as an indicator of mobility opportunities (cf. e.g., Cerwenka, 1982: 30 ff; Altshuler, 1979). Indirectly, the different mobility opportunities of ethnic and socioeconomic groups can also be inferred by examining who benefits from increased mobility by settling in places with higher residential quality (Nelson/Clark 1976; cf. also below).

However, these analytical distinctions should not hide the fact that there is a strong interdependence between mobility patterns and modes of transport. For example, the structural organization of the transport system has a major influence on mobility behavior. The mere existence of a new means of transport (rail network, timetable, availability of the car, roads, and motorways) increases individual options (mobility behavior) with regard to the type of transport and the accessibility of places (e.g., Jones 1987: 41). Increased accessibility of a place in turn makes residential and industrial settlements more attractive. For example, as employers, retailers, and citizens adapt to the new conditions, the spatial distribution of activity locations (mobility structure) can change in the longer term, as outlined above. However, a new mobility structure can also force mobility from the outset (Linder *et al.* 1975) and exclude certain modes of transport (bicycle, feet). This ultimately influences the mobility opportunities of groups who cannot drive or travel by car or train (children, poor people). They are *even less mobile* than before as a result of changes toward mobility structures that are more distance intensive.

This work focuses on the interrelationships between mobility patterns and motorized modes of transport in Los Angeles in their historical course. This topic will be dealt with in two steps. In the first step, a *historical analysis* will present the transport policy development in

Los Angeles since the end of the 19th century and describe the emerging mobility patterns. In this context, it will be discussed, with which strategies the political actors reacted to the mobility problems of the metropolis and how the strategies affected in turn the mobility structures. Of particular interest is the reorientation of transport policy in Los Angeles in recent years from the construction of a freeway network to the construction of a rapid transit system.

In the second step, elements of the development of a transport policy in Los Angeles will then be *explained based on* various factors. The aim is to provide a deeper understanding of the relationships between transport modes, mobility patterns, and the transport policy. While political–economic factors for the decline of the railways will be examined first, the transport policy action patterns will then be derived from interests and institutional framework conditions. Finally, sociocultural variables will be used to discuss the relatively permanent orientations of political and social actors that have been decisive for the development of the transport policy and the formation of specific mobility patterns.

From a methodological point of view, the metropolis of Los Angeles is a stroke of luck since, at various periods of time, first the light railways and then the automobiles held a monopoly-like position as the mode of transport. This makes it possible to track the effects of the individual motorized modes of transport separately over time. The analysis, which covers a period of about a century, is methodologically based largely on document analysis. In addition, in-depth interviews were conducted with transport policymakers, representatives of interest groups, and scientists of the metropolis.

In the following Chapters — 2, 3, and 4 — the development of the transport policy in Los Angeles is described, with particular reference to mobility patterns. In accordance with the chronological sequence, Chapter 2 focuses on trams and regional trains, which were the dominant modes of transport in the metropolis until the end of the First World War. Since the 1920s, the automobile assumed a monopoly-like position (Chapter 3). Chapter 4 deals with the return of the railway as a mode of transport in the metropolis in 1990. Chapter 5 will then compare the mobility patterns and dominant transport policies of the different periods. Chapter 6 discusses factors that explain development aspects of the transport policy in Los Angeles. With the new edition of this book, Chapter 7 was added, in which the traffic and transport trends in Los Angeles between 1990 and 2022 were analyzed.

Chapter 2

Road and Regional Railways as Dominant Modes of Transport

It would never do for an electric line to wait until the demand for it came. It must anticipate the growth of communities and be there when the homebuilders arrive — or they are likely not to arrive at all, but to go to some other section already provided with arteries of traffic.

Henry Huntington, 1904

The growth of Los Angeles from a small city to one of the most populous metropolises in the US is inextricably linked to the advent of trams and regional trains at the end of the 19th century. Although the railways were the dominant mode of transport in the metropolis from 1890 to about 1920, they had an extraordinarily lasting influence on mobility patterns. Within a short period of time, the railways distributed the rapidly growing population across the metropolis and created the horizontal settlement structure that is characteristic of Los Angeles today. In the following chapter, the emergence and decline of the railways as well as the interrelationships with mobility patterns are described. Public action has been of little significance in this phase. Rather private actors exerted great influence on the development of the transport policy. The reasons for the reluctance of political decision-makers to act will be discussed later (cf. Section 6.3.1).

2.1 Urban development and railways

Although settled as early as 1781, Los Angeles remained a small, relatively insignificant village until the 1870s, with a population of around
6,000. Within a few years, the population of L.A. multiplied, so that by
1890 the city already had 50,000 inhabitants. Thirty years later, the greater
Los Angeles area was already home to over a million people. Between
1890 and 1920 the metropolis thus had the highest growth rates in the
whole country (Brodsly 1981: 72).[1] As early as 1930, the city passed the
one million mark and the county passed the two million mark (cf.
Table 2.1).

The initial spark for the metropolis's population explosion was the
connection to the transcontinental railway network with the completion of
the railway line from San Francisco to Los Angeles in 1876 by the

Table 2.1:   Population development in the metropolis of Los Angeles (in thousands).

Year	Los Angeles City	Los Angeles County	Counties (except Los Angeles County)	Metropolis (CMSA)
1870	6	15	n.a.	n.a.
1880	11	33	n.a.	n.a.
1890	50	100	n.a.	n.a.
1900	102	170	67	237
1910	319	504	120	624
1920	577	936	181	1,117
1930	1,238	2,208	333	2,541
1940	1,497	2,785	393	3,197
1950	1,970	4,152	665	4,817
1960	2,479	6,038	1,501	7,539
1970	2,816	7,032	2,627	9,659
1980	2,967	7,476	3,588	11,064
1990	3,484	8,863	5,669	14,532

Source: US Bureau of the Census (population), various years.

---

[1] This population explosion was not limited to Southern California, but was the expression
of a general rural exodus in the United States (Brodsly 1981: 73 f).

Southern Pacific Railroad. Five years later, the same operator built another link, connecting the city directly to the East. In addition, the *Santa Fe Railroads* opened a competing line in 1885, creating the conditions for L.A. to become the urban center of the Southern California region (Brodsly 1981: 61–63).[2]

The extreme population growth coincided with the technical maturity of modern urban transport. Although the first means of mass transportation in Los Angeles, the *Pioneer Omnibus Street Line*, was still pulled by horses in 1873, the development from *horse cars* via *cable cars*[3] to electric *streetcars* progressed rapidly. Although L.A. had less than 50,000 inhabitants at the time, it was one of the first cities in the US to introduce both cable cars and electric trams. L.A.'s progress was evident not only in the existence of two *modern* modes of transport but also in the size of the network. In comparison to Philadelphia (1,047,000 inhabitants) and Boston (448,000 inhabitants), which at that time were still completely dependent on horse cars, Los Angeles already had a respectable network of cable cars and trams, which supplied the population surprisingly well (Nelson 1983: 265).

Trams and regional railways had a significant impact on mobility patterns, especially in the short period from 1890 to 1910. During these twenty years, not only did the city's population rise from about 50,000 to 320,000 but in Los Angeles County alone the population tripled from 170,000 to 504,000 between 1900 and 1910, producing a significant demand for mobility. At the turn of the century, the electric tram network was considerably expanded in all directions. By 1898, the Los Angeles Railway already had 103 trams running on a network of 72 miles. The lines all ran from downtown Los Angeles to the surrounding area (Eagle Rock, East L.A., Boyle Heights, Vemon, Inglewood, and Pico Heights). By 1911, the company already had 525 wagons running on over 172 miles of double track and 350 miles of single track (Nelson 1983: 266; Fogelson 1967: 92).

---

[2] Paradoxically, the relative insignificance of Los Angeles was its greatest advantage. As Brodsly suspects, the small town with its insignificant harbor, in contrast to San Diego further south, did not compete with San Francisco, which was the heart of the railway companies, on which their money mainly depended (Brodsly 1981: 61 ff).

[3] The cable cars that can still be admired in San Francisco today are driven by a cable moving at a certain speed, which runs underneath the road. By means of a clamp, the cable car can grasp the cable through a crack in the road and thus set the car in motion.

From the large number of transport companies, two large operators prevailed until 1911: on the one hand the *Los Angeles Railway Company* (LARY), under which all trams were united, with its sole owner Henry Huntington, and on the other hand the *Pacific Electric* (PE), under which all regional lines were merged. The new owner of LARY and PE was the Southern Pacific Railroad, which was said to have had less interest in passenger services than in the more lucrative freight services (Nelson 1983: 268). Nevertheless, unlike many other cities in the US, the inefficient competition of many local transport companies operating in the same areas was eliminated early on.

The importance of rail companies for urban development cannot be understood solely from their transport business. The relatively low revenues from passenger volumes would often not justify the capital requirements for the construction of many lines and operating costs for the same, especially in the case of PE (Fogelson 1967: 104). Rather, there was a linkage with economically more profitable services. The *combination of passenger transport and real estate trading* — often within a single company — allowed for large profits. It became apparent early on that, depending on where certain railway lines were laid, the adjacent real estate assets increased considerably.

An early example of this was Robert M. Widney, who in 1874 laid the railway lines for the city's first horse-drawn tram right by his house, only to have it end a few streets away at a property that was also in his possession. Soon, the commercial center developed around this track, which caused land prices to soar (Nelson 1983: 264). The intertwining of passenger transport and real estate speculation was obvious:

> Between 1880 and 1910, cable cars and electric trolley lines were built
> by holders of large tracts of vacant land with the specific intention of
> subdividing that land and profiting from the sale of homesites made
> accessible to downtown by transit (Wachs 1984: 300).

The classic example of this practice was Henry Huntington. In addition to his activities as a railway entrepreneur, he was also active in the real estate business as one of the largest landowners in the region. His strategy was to purchase large areas of undeveloped land that seemed suitable for his development plans and then to lay a railway line between this undeveloped land and downtown Los Angeles. Examples of this approach are the construction of the lines to Monrovia, South Pasadena, and San Marino (Brodsly 1981: 71). The construction not only led to the expansion of already existing small settlements but also led to the development of no less than 13 new towns that were built with his personal support as mentioned by Brodsly. All of them, except one, were located directly on the network of his regional trains (*ibid.*).

In contrast to the large cities on the East Coast, where modes of transport followed the growth of the population and could therefore expect a secure volume of traffic, in Los Angeles, the opposite development took place. Individual profit expectations from land development led the investment behavior of rail and real estate companies. The expectation of prompt profits from the sale of land and the hope of later income from the railway lines, for which the right of way was secured through long-term concessions, were the motives for the railway and real estate companies to open up the metropolis. While the rail companies of other large cities of that period were largely only *meeting* the transport needs of their inhabitants, the strategy in Los Angeles was to *create* transport needs by providing good transport services in the greater area.

2.2 Mobility patterns

While Los Angeles was still a small city of 6,000 inhabitants in 1870, where the places of activity could be reached on foot, a high volume of traffic was created by the combination of population explosion, real estate speculation, and railways. The effects of the railways on mobility patterns will now be discussed in detail. In the following two sections, we will first analyze the impact of railways on the mobility structure. To this end, the spatial location of the Angelenos' places of residence (settlement patterns) (2.2.1) and the locations of work and consumption activities are examined (2.2.2).[4] Subsequently, the impact of this structure on the mobility behavior of the Angelenos will be examined. Finally, the mobility opportunities of the Angelenos will be outlined using ethnic and socioeconomic groups as examples.

---

[4]The overwhelming majority of recreational sites will not be dealt with in this context as there was insufficient material available for this purpose. However, it can be assumed that during the *railway phase* of Los Angeles' transport policy development, a majority of leisure venues (such as cinemas and restaurants) were located in the center of the city.

2.2.1 Railways and urban sprawl

The population explosion created a huge demand for housing. The trams and regional trains, together with real estate companies, made the offer for people to live in the metropolitan area, which was eagerly taken up by the masses. In many regions of the metropolis, the connection to the electric railway network had a direct effect: The population of Long Beach, a community 20 miles south of the city center, grew from about 2,000 to nearly 18,000 people in the first decade of the twentieth century, following the construction of a suburban railway line. Similarly, Hollywood, northwest of the Central Business District (hereinafter CBD), increased its population from 600 to 10,000 between 1903 and 1910. Out of 42 cities that were incorporated into the urban area of Los Angeles in the mid-1930s, 39 owe their early growth to regional railways (Brodsly 1981: 68 f). In Los Angeles County, about 100 cities were founded in anticipation of the railway in the 1880s alone.

During this period, many cities developed into pure *dormitories*, which in turn required good transport connections. In fact, the frequency of trains was very high. Between downtown L.A. and Long Beach, for example, more than 100 trains ran daily, every 15 minutes (Nelson 1983: 267). Watts, located halfway between downtown L.A. and Long Beach on the intercity line, became "the first full-fledged bedroom community in the city. Supporting only minimal commercial activity itself, it became one of the Southland's most important centres in the electric rail system" (Adler 1966: 22 ff, quoted after: Brodsly 1981: 68).

In advertising for newly developed areas, many real estate companies emphasized the immediate vicinity of the railways, which provided quick access to the CBD:

> These magnificent lots are located at the end of the Washington Streetcar Line, in our Westview Terrace tract, only twenty minutes from downtown, with a 5 Cent fare and car service every few minutes (*Los Angeles Express*, 7.12.1912).

The tramways, and especially the regional railways, have shaped the sprawling structure that is still typical of Los Angeles today. The main argument in favor of this theory is the fact that, in addition to the central areas of the city, the surrounding areas also showed considerable population growth rates. As Table 2.2 shows, the population growth between

Table 2.2:   Population of the metropolis of Los Angeles by region 1900–1920.[a]

Region	1900	1910	1900–1910 (%)	1920	1910–1920 (%)
Valley	5,846	17,616	201.3	32,869	86.6
East	31,598	80,042	153.3	132,901	66
Central	102,479	313,104	205.5	568,886	81.7
West	8,536	24,517	187.2	65,904	168.8
Southeast	32,461	72,020	121.9	151,055	109.7

Note: [a]Pegrum counts only the counties of Los Angeles and Orange in the metropolis of Los Angeles. The five main areas into which Pegrum divides the city comprise the following subdistricts: Valley includes the San Fernando Valley and Glendale; East includes Pasadena, Pomona-Foothill, and Alhambra; Central includes Northeast, East, Central (including CBD), Wilshire, and Hollywood; West includes Beverly Hills-Westwood, Santa Monica Bay, and Adams-Inglewood; Southeast includes the Orange County Southeast, Whittier-Norwalk, and Southeast (see Pegrum 1968: 564). *Source*: Pegrum (1968: 564).

1900 and 1920 was high, especially so for areas far from the center of the city.

At the end of the First World War, the regions of the Valley, East, and Southeast were already home to almost half of the metropolis's population, even if — in reference to Pegrum (1968: 564) — the metropolis only consisted of the Los Angeles and Orange Counties and the center of the city was defined very broadly. It then also consisted of Hollywood and Wilshire, areas that lay miles away from the CBD.

The central area of the town, which was mainly served by the LARY, accommodated a majority of the population in a comparatively small area. However, the extension of the PE rail network, with over 1,164 (single track) route miles *in four counties* and over a hundred miles from one end to the other (1923), allowed a large part of the population to live in the sparsely populated suburbs of the region. Spencer Crump, in his description of the regional railways, emphasized the extensive network of routes:

The Big Red Cars, traveling on right-of-ways lined by mile after mile of waxy-green orange groves, went through Covina, Claremont, Upland, Etiwanda, and Rialto en route San Bemadino, Redlands, and Riverside — more than sixty miles from Los Angeles (Crump, zitiert in: Richmond 1991: 27).

Los Angeles operated the largest rail network in the world at that time (Bottles 1987: 31), despite the very low population density in the metropolis.[5]

Figure 2.1 shows not only the wide extent of the network but also the radial structure of the railway lines radiating from the center. The horizontal structure of the city is a consequence of the regional railways, which spread the population far across the county, especially during the first two decades of the 20th century (Wachs 1984: 300).

While one can relate suburbs on the East Coast to an older urban area, allowing one to escape from a dense urban core, this is hardly meaningful in Los Angeles, as a dense core interspersed with housing never existed:

> Residential patterns in Los Angeles (…) were commonly all suburban.
> It was a city of single- and two-family detached homes, a category, that
> as late as 1930, comprised 93 per cent of all dwellings (Brodsly
> 1981:76).

In Chicago, by comparison, the proportion of single-family homes was 52% and in Boston only 49.5% (Fogelson 1967: 146).

Apart from the railways being the most important component, other factors also encouraged the tendency toward a horizontal settlement structure:

- The inflationary development of suburbs, or the lack of a city center with skyscrapers and high population density, was supported by the ban on the construction of tall buildings. After the great earthquake of 1906, which destroyed large parts of San Francisco, such a law was passed and remained in force until the 1950s. This reduced the attractiveness of the CBD for large companies, as tall, prestigious buildings — due to expensive land prices for a small area — were not possible. The only exception to this ban on tall buildings was the *City Hall*, completed in 1928.

---

[5] Even after the considerable population boom of the 1920s — partly due to the car — the population density in the entire urban area of the metropolis in 1930 was 1572 inhabitants per square mile. In the metropolis of New York it was 4336, in Chicago 3890, and in Philadelphia 2868 inhabitants per square mile. The density outside the city center in LA was only 1045 inhabitants per square mile (Fogelson 1967: 143).

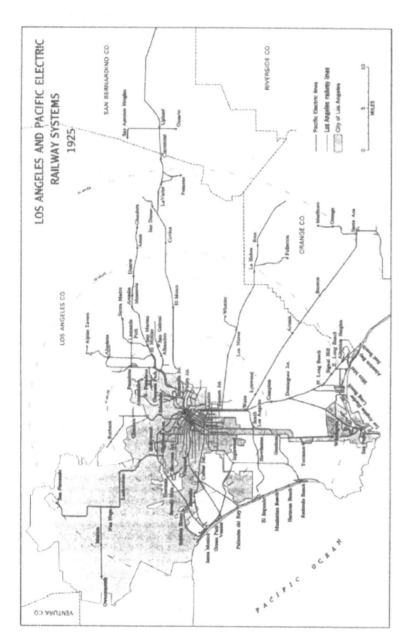

Figure 2.1: **The electric railway network of the metropolis of Los Angeles in 1925.**

Source: Fogelson (1967: 173).

- Another factor for urban sprawl is the relatively late importance given to the Port of Los Angeles, which is located about 20 miles from the downtown area. In contrast to other cities where the port was the source of wealth around which the city developed (e.g., New York), the port of Los Angeles only began to play an increasingly important role with the boom of the 1920s. This was due to the discovery of seven large oil fields (Nelson 1983: 181) and the need to meet the growing demand for oil by ship.[6] As a result, the area around the port gained in importance (the port city of Long Beach is — after Los Angeles — still the second largest independent city in the metropolis today), which in turn forced increased communication with the CBD. The great distance between the two economic activity locations had to be covered by transport systems (rail, then road), which naturally made the space in between more attractive for settlements.

Of course, the layout of the railways determined the settlement structure relatively rigidly. Since no other motorized means of local transport was available at that time, the Angelenos, who lived in the suburbs, were heavily dependent on the mobility that the railways provided. So, Foster emphasizes the following:

> Pre-World War I residents were so dependent upon the trolley for transportation that developers made few attempts to promote single-family homesites more than a half-mile from the lines (1976: 476).

Even city plans drawn up in 1919 show only few streets more than five or six blocks away from the tram and regional railway lines (Foster 1976: 476; Bottles 1987: 183).

2.2.2 Jobs and consumer locations

L.A.'s light railways thus contributed significantly to the urban sprawl of the residential areas. Their influence on the spatial location of other places

---

[6]The importance of the port, which only began to grow in the 1920s, is reflected in the development of cargo turnover. While in the fiscal year 1920, 2,886 ships with 3.5 million tons of freight reached the port of L.A., ten years later there were already 8,633 ships with 26 million tons of freight (see Wachs 1984: 303).

of activity (workplaces, places of consumption, etc.) is discussed below. The mobility structures then represent the main factors determining the mobility needs of the population. The further apart the activity locations are from each other, the greater the demand for motorized transport.

A characteristic feature of Los Angeles is the *strict separation of housing and work*, which, incidentally, was also preferred by the population (Fogelson 1967: 139 ff; cf. also Section 6.3). In order to make the suburbs more attractive to the population as residential areas, the *developers* (mostly private land development companies) prohibited the settlement of trade and industry in most newly developed areas. "In short, deed restrictions were employed by the subdividers to ensure that most of greater Los Angeles' suburbs would stay strictly homogeneous and purely residential" (Fogelson 1967: 146).

Indeed, until the 1920s, a majority of jobs were concentrated in or near the CBD (Pegrum 1968: 560; cf. also Soy/Morales/Wolff 1983: 207 ff). Only the area around the harbor, in the south of the city, and the developing film industry in Hollywood offered some jobs outside downtown L.A. at that time. The construction of large banking and commercial complexes, utilities, and department shops made the CBD the "most concentrated section in southern California" (Fogelson 1967: 148). In 1920, the CBD still held "more than three-quarters of Los Angeles' commercial and professional enterprise" (*ibid.*). Even in 1929, about three quarters of all department shop sales in Los Angeles County were still made within the CBD (Bottles 1987: 194 f). A transport study from 1924 also confirms the economic position of the CBD. Thus, 1.2 million people (i.e., more than the entire population of the city) entered and left this sector every day (Fogelson 1967: 147).

Two factors in particular prompted industrial and commercial enterprises to locate mainly in the center of the city, even though a large part of the population already lived in the surrounding area. First, the population was too scattered in the outskirts, so that the potential demand there, for example, for retailers, was too low (Fogelson 1967: 151). On the other hand, the lines of the railways, all of which were laid out radially toward the center (cf. Figure 2.1), ensured that customers had easy and cheap access to the CBD. The companies had no problems attracting customers. They therefore lacked the incentive to change their business locations.

Second, there was no satisfactory means of transporting goods in the surrounding area. Although the electric railways successfully transported the population to and from their homes outside the center, they were not suitable for freight transport.[7] Rather, industry and commerce in the first two decades of the 20th century were still dependent on other inefficient means of freight transport, mostly horse transport (Brodsly 1981: 69). The motorized lorry, or "truck", only became established after the First World War. However, it then very quickly allowed companies to decentralize.[8] Without adequate means of transport, however, the economic locations concentrated on the main traffic junctions: freight stations in the city center and shipping piers in the port.

The majority of jobs — with the exception of local supply — thus remained in downtown L.A. It was not until the boom of the 1920s that trade and industry became increasingly decentralized, although the dispersal of residential areas increased considerably with the advent of the automobile (Wachs 1984: 302; cf. below: Chapter 3).

2.2.3 Mobility behavior

In the following, the effect of the mobility structure formed by the railways in essential parts on individual mobility behavior is examined. For the first two decades of the 20th century, however, little systematic information can be found on Los Angeles. For example, no studies have been carried out to determine the average journey or route lengths, travel times, and travel rates of citizens per day. The only useful data were provided by the railway companies. The annual transport figures underscore their dominance as the metropolis's motorized modes of transport. The volume of transport in relation to the population indicates a comparatively high number of trips. Bottles calculates that between 1910 and 1915 the average Angeleno made over 500 journeys a year on trams and regional trains (1987: 266). This means that more than 1.3 trips per capita (including

---

[7]For example, the transcontinental railways relevant for freight transport (the PE transported very little freight) were only able to supply trade and industry with goods in some parts of the city. In contrast, the city center was very well served by the railways (cf. Fogelson 1967: 151).

[8]The importance of the truck, the use of which only began after the First World War, is shown by the registration figures: In the entire United States, only 10,100 trucks were registered in 1910, compared with over 1.1 million in 1920 and as many as 3.6 million 10 years later (Motor Vehicle Facts and Figures 1977).

children, the elderly, and the sick) were made by tram every day. In other cities of similar size, this figure is only half as high for the time (Bottles 1987: 33).[9]

However, since 1913, the annual per capita use of the railways tended to decline. While the PE *continuously* lost passenger shares, the LARY was able to increase its passenger volumes per capita again until 1923. After that, however, the number of trips per capita also decreased rapidly for the LARY. While the first slump was caused by the influence of the *jitneys*[10] (which, however, as a mode of transport soon disappeared), after the First World War, the automobile entered into competition with the railways and replaced them as the dominant mode of transport in the 1920s (cf. Chapter 3 and, for explanation, Section 6.1).

It can be stated that the Angelenos used the railways as their dominant means of transport until the end of the First World War. However, due to a lack of surveys, further data on the mobility behavior of the Angelenos in the railway phase cannot be directly accessed. The mobility structures, which are intensive in terms of distance, due to the low settlement density and the strong spatial separation of the activity locations in the metropolis, however, indicate, long distances. Moreover, until the end of the 1920s, most department shops were located in the CBD, from which further (distance-intensive) consumer trips could be derived. The fact that the Angelenos used the railways twice as often as residents of other cities points in the same direction. It is not possible to determine to what extent the long distances usually covered by the railways also had an impact on the daily travel budget (the time spent on reaching the places of activity in traffic). However, it is obvious that the Angelenos also had to spend more time in traffic than inhabitants of other cities due to the long distances involved.

---

[9] In the 1990s, only the "tram city" of Zurich came close to this high figure of 500 public transport journeys per capita and year.

[10] The private taxis, which were quickly called "jitneys" by the local population, appeared on the streets of the city in 1913. The vehicles running on the main roads robbed railway companies of the valuable and numerically large group of short-distance passengers. In 1915, the 1,800 jitneys carried about 150,000 passengers daily. When they were also subject to taxes and charges in 1918, their operation became unprofitable and they quickly disappeared from the roads (Fogelson 1967: 166 f).

2.2.4 Opportunities for mobility

Since little empirical information can be drawn on for this period, only very limited statements can be made about the mobility opportunities of various groups in a "rail-mobile" society. However, it can be plausibly argued that poorer sections of the population benefited less from the large railway network of the metropolis. In particular, richer families were able to use the railway lines to escape the narrowness and dirt of the city center.

Since a large proportion of newcomers were very wealthy compared to other cities, more citizens in Los Angeles were able to take advantage of the mobility offered to settle in small suburbs. As Brodsly emphasizes, "South California was attracting the social cream of the East and the midwest, retiring farmers, businessmen and professionals" (1981: 64). "(T)he outstanding quality of the newcomers was their prosperity," noted a San Francisco newspaper, "(t)hey are almost invariably persons of American birth, good education, and some means. (...) This is the best American stock; the bone and sinew of the nation, the flower of the American people" (quoted in: *ibid.*: 62 f). Southern California, with its Mediterranean climate, also attracted rich elderly and sick people who preferred to retire there. Between 1890 and 1940, for example, only 20% of the citizens of Los Angeles County were foreign-born or non-white, while this proportion was twice as high in Chicago or New York (Brodsly 1981: 77).

Los Angeles was thus home to *only a few poorer social groups* at that time, mostly Mexicans, US Mexicans, and blacks (cf. Hirsch 1971: 8 ff). However, these groups were hardly able to benefit from the spatial mobility gains of the railways to the same extent as the richer white part of the population.

The increased immigration of blacks after 1910, who until then represented only 2% of the population, alarmed the Anglo-American population and acted as an impetus for many to move away from their old environment in central L.A. to the suburban communities. In many places, blacks moved — usually as tenants — into the vacated houses, although even then, most newcomers settled in South Central Los Angeles, just south of the CBD. The white strata of the population vehemently resisted *mixing* the suburbs with other ethnic groups: "Whites resisted black expansion into the suburbs through physical intimidation and the use of restrictive covenants forbidding the sale of property to non-whites" (Bottles 1987: 181 f).

Similarly, the expansion of the largest minority in Southern California, the Latinos, into white areas was prevented: "As Los Angeles' Mexican population rose throughout the twenties, whites successfully increased their efforts to stop Hispanic residential movement into the city's northern and western suburbs" (*ibid.*: 182). In order to meet the wishes of future suburban residents, who were generally recruited from among the richer, white citizens, the land development companies laid down documented housing restrictions that were to guarantee homogeneous settlement:

> These (…) prohibited occupancy by Negroes and other Orientals in most tracts and, in the more exclusive ones, fixed minimum costs for houses so as — in one developer's words — 'to group the people of more or less like income together' (Fogelson 1967: 145).

For example, the building regulations (fixed by the entrepreneurs) in newly developed areas often only allowed very specific types of houses that only the respective income groups could afford.

The chances of traveling long distances to live in *better* housing areas were therefore largely limited to WASPs, i.e., white Anglo-Saxon Protestants, middle and upper income groups. Even then, land use and planning regulations as well as simple expulsions led to the ghettoization of Mexicans, Japanese, and blacks. They had to make do with the poorer housing in or near the CBD and also in the port area. The extensive network of railways theoretically offered everyone the chance to live in the beautiful suburbs and still get to the other places of activity. In fact, however, only the better-off socioeconomic groups benefited from this. The *use of the railways* was therefore *selective* in that the trains transported mainly white upper- and upper-middle-class citizens to the suburban communities.

Another element of "rail mobility" in Los Angeles is the paradox that, on the one hand, it made the different areas of the city more accessible, i.e., it connected spaces, and, on the other hand, the closer spatiotemporal connection led to social separation and spatial segregation of people. In this sense, the railways can be described as a welcome instrument which — in addition to the *socioeconomic* separation of different ethnic and social strata — also made *spatial* distance possible. Escape from undesirable social strata was at least partially feasible. However, since jobs and places of consumption were located almost exclusively in the center of the city and the railways were the dominant means of transport, contact with other social groups, and the problems of the city in the truest sense of the word, was still inevitable.

2.3 Mobility problems of local rail transport

Light railways and public transport are often categorically claimed as solutions to congestion problems. In Los Angeles, on the other hand, it was shown that rail-bound local transport, as the dominant mode of transport, was itself already frequently congested at the turn of the century. It follows that mass transport cannot solve mobility problems *per se*. It is also helpful for the analysis of mobility problems to focus on the interrelationship between modes of transport and mobility patterns.

Congestion during peak traffic hours was already identified as the number one problem in major cities in the US in 1905 (Owen 1966: 6). In Los Angeles, the situation worsened considerably in 1911. The *Los Angeles Examiner* reported the following:

> There are times in the rush hours when every foot of trackage in the business district is covered with trolley cars (…) It (takes) more time to get from the Station at Sixth and Main to Aliso Street than it (does) to ran the balance of the trip to Pasadena (17.11.1911).[11]

The first report of the Board of Public Utilities in 1911 described the situation in public transport as "chaotic" and noted that "the congestion of street car traffic in the business district at the rush hour is indefensible" (quoted after: Bottles 1987: 45). Bion J. Arnold, the traffic expert appointed by the City Council after the Committee's investigations, provided the following figures for the generally perceived congestion problems:

> Fully 40.000 riders on both Systems (i.e., LARY und PE; annot. of author) are delayed from five to forty minutes during the rush hours each

---

[11] While Main and Aliso Streets are not far apart in downtown, Pasadena is about 15 miles to the northeast of the city. The Los Angeles Times (25.7.1909) and the Automobile Club of Southern California (1922: A Report on Los Angeles Traffic Problems with Recommendations for Relief, Los Angeles) also report traffic jams and delays in the downtown area.

day, and as many are inconvenienced during the non rush hours due to fundamental defects of the transportation arrangements along Main Street (*ibid.*).

What led to these traffic and transport problems? According to Hägerstrand, the personal interaction of people requires the bundling of paths at a destination. If many people have the same destination at the same time (regardless of whether this is personally desired), then, depending on the type of locomotion (or means of transport) and the available space of the respective destination route and destination, there is a compression that can lead to congestion, i.e., a standstill of traffic (Hägerstrand 1987: 13 ff).

In Los Angeles in the first two decades of the 20th century, the predominant means of transport was the tram and regional train, and the destination was the center of the city, where the paths of the citizens involuntarily converged. The mobility structure characterized by a sharp division of uses (living outside — work within the city center) forced long distances and caused a high temporal and spatial concentration of traffic flows, which led to congestion during rush hours.

Two factors also supported the compaction in Los Angeles. First, Los Angeles' CBD had a very dense road network at that time compared to other cities. While streets took up 29% of the city center in Chicago, 39% in Cleveland, and as much as 44% in Washington, L.A. devoted only 21.5% of its area to roads (cf. Brodsly 1981: 85). As one city planner rebuked, "There are surprisingly few streets of generous width in Los Angeles. (…) The prevailing standard has been a sixty foot street, a width totally unsuited for traffic street of great capacity" (*ibid.*: 84). Brodsly even reported that few streets were more than 30 or 40 feet wide (= approx. 9 or 12 meters) (*ibid.*).

However, as explained, the respective limited space was only one factor among several. On the other hand, the meeting of the two rail networks in the city center also contributed to congestion and delays. This was a consequence of the radial routes and the importance of downtown as a business and industrial district (mobility structure). The closer the regional trains of PE (which operated outside the center on private tracks) got to the center the slower they became. Until they reached their central terminal in the middle of the CBD (6th/Main St.), they additionally hindered the operation of the LARY trams. Congestion was further exacerbated by the fact that almost all other routes of the Angelenos passed through downtown L.A., as direct connections between suburbs were practically nonexistent. The low population density in the region made the construction of tangential rail links between the suburbs unprofitable for rail operators. In order to get from one nearby suburb to the next, for example, citizens were therefore forced to take a detour to the center to change trains. The high travel length and mobility time budget of the citizens is partly explained by these additional journeys, which made the railways more and more unattractive with the increasing importance of the region.

2.4   The decline of tram and regional railways

The population boom of the 1920s brought the electric railways of the metropolis their highest transport volume. LARY and PE together carried 356.5 million passengers on the world's largest electric railway network in 1924. The LARY slipped back into profit by the beginning of the depression and the PE was also able to make a profit in 1923 (see Appendix).

However, these passenger gains disguise the fact that the railways had long since lost their transport monopoly. According to a traffic census, 48% of citizens traveled to the CBD by car as early as 1923. Eight years later, the figure had risen to 62% (Bottles 1987: 204). On top of this, the cars drew passengers away from the railways during the *cheap* transport service outside peak hours and during weekends, without relieving them of the expensive and unprofitable rush hour traffic. PE alone had to handle 50% of the passenger volume on working days for two hours in the morning and two in the afternoon (Fogelson 1967: 179 ff).

Losses in previous years prevented the railways from expanding their network of routes, inter alia through investment. The population boom of the 1920s, together with the new private transport mode, led to a considerable expansion of the settlement area, which was no longer served by the railways. The attractiveness of the railways was also considerably reduced by the congestion problems of the 1920s. The traffic in the center, which had become considerably denser due to cars and lorries, increasingly came to a complete standstill, so that the railways in the CBD made even slower progress than before.[12]

---

[12]"The Times reported that streetcars 'full of passengers hanging onto straps and clinging to rear fenders' required thirty minutes to move six blocks in the downtown area during the rush hour" (Bottles 1987: 125).

The assumption of many that the change of citizens' mode of transport to the car would reach a saturation point with the deterioration of the traffic situation in the city center turned out to be completely wrong. It is true that the pressure of problems, in the form of traffic jams and delays on the routes leading from the surrounding areas to the CBD, was increasing. But instead of using public transport again, citizens increasingly avoided the congested area and surrounded the city center with their cars. As late as 1923, 68% of citizens living within a ten-mile radius of the CBD commuted to the city center, compared to 52% in 1931 (Baker 1933: 37d).

With the increasing inaccessibility of the CBD and the improvement of transport facilities, trade and industry began to relocate from the center to other regions at the end of the 1920s, further worsening the competitive position of the railways. "And once the automobile not only competed with the railway but also established the setting for competition, the future of urban transportation rested with the motorcar as the past had with the train" (Fogelson 1967: 180).

As late as 1925, the California Railroad Commission published a report recognizing the serious situation of the LARY and the PE (Fogelson 1967: 172). In its recommendations, it urged L.A.'s city council to help the ailing railways. In order to ensure more efficient operation and to avoid competing lines, the Commission concluded that it would make sense to transfer the two private companies to municipal ownership. The Commission argued that this would remove, for example, the 5.5% turnover tax paid by the state and reduce the high interest rates on bonds issued by the railways. In fact, New York and San Francisco, for example, were already operating urban transport companies at that time. In Los Angeles, too, the municipality took over the electricity and water supply and the port at the beginning of the 1900s.

The negotiations with Huntington's LARY almost led to an acceptable result. The electric tram would have secured much needed investment and the city would have purchased a public transport system at a fair price of $30 million. However, the death of LARY's owner Henry Huntington prevented the negotiated contract from being signed, as his heirs were not interested in selling.

The situation was even worse in the mid-1920s for the PE, which did not take part in these negotiations. Despite the considerable growth and spatial expansion of the population, PE's lines in 1925 extended over almost the same area as ten years earlier. With the exception of a small profit in 1923, the PE always led in the red. Its parent company, Southern

Pacific Railroad, financed the operating costs but not the necessary investments for line extensions, renewal, and improvement of the outdated
working capital (Fogelson 1967: 175).[13]

The Great Depression, which began in 1929, worsened the situation
of the railway companies. The sharp decline in the volume of traffic had a
devastating effect on the financial situation of the railways (see Annex 2).
In 1932, 70% of railway companies' working capital was considered outdated (Nelson 1983: 272). The companies had no choice but to abandon
their unprofitable lines or replace them with buses, reduce journeys, and
increase ticket prices. This of course did not make the city's public transport more attractive.

Whereas in 1926 only 15% of PE's passenger miles were covered by
buses, this figure rose to 35% in 1939. The total route length of PE's buses
had already exceeded that of the company's railways three years earlier
(Richmond 1991: 36 f) With the exception of the Second World War, PE
continued to make losses. Rubber and petrol rationing caused by the war
led to a short-term increase in the number of journeys by regional railways. After the Second World War, the annual loss was again $2.2 million,
and a year later it rose to $3.4 million (Richmond 1991: 38).

The substitution of railways by buses continued after the Second
World War as soon as the Railroad Commission granted permission
(Brodsly 1981: 93). In 1953, Southern Pacific sold its subsidiary PE to
Metropolitan Coach Line, which operated mainly buses in Southern
California. The company made only losses with PE during its five years
of ownership and ceased operating two railway lines. In 1958, the *Los
Angeles Metropolitan Transit Authority* (MTA) purchased the three
remaining intercity lines of PE together with the last five tram lines of
LARY. On 8 April 1961, the last line of the "Big Red Cars" was finally
discontinued (Richmond 1991: 39).

The same happened to LARY, whose owner, the Huntington Estate,
sold the railway to American City Lines in 1944. The latter was a subsidiary of National City Lines (NCL), which bought railways throughout the
United States and replaced them with buses. General Motors, together
with Standard Oil of California and Firestone Tires, among others, was a

---

[13] Of a planned underground system in the city center, PE, in view of its precarious financial situation, only completed a one-mile-long tunnel, which was opened in 1925. This
tunnel reduced the travel time of trains from the north and west by a quarter of an hour.
But given the changing situation in the city, this was no more than a drop in the ocean.

shareholder in NCL (See Section 6.1.1). The federal *antitrust law* forced American City Lines to sell the network, renamed Los Angeles Transit Lines, to the MTA. The MTA operated the remaining five tram lines until it also replaced them with buses in 1963 (Richmond 1991: 39).

In the early 1930s, there were attempts to agree on a comprehensive high-speed rail network in the face of a strong outflow of shops from the center. However, the construction of such a railway system could not be financed privately. Even New York, a city with a much higher population density, was making losses and claiming public subsidies every year to support operating costs (Bottles 1987: 163). In Los Angeles, public financing of a railway network was no longer feasible at that time. As early as the mid-1920s, citizens clearly rejected the financing of a rapid transit network in a referendum and voted for road transport (cf. Section 6.2). The overwhelming majority of the population had already clearly decided in favor of the car.

Chapter 3

Los Angeles as an Extreme Form of an Automotive Conurbation

We shall solve the city problem by leaving the city.

Henry Ford

Transport policy developments have so far been described largely without reference to public action. In the first phase, the development of transport modes and mobility patterns was mainly determined by the privately organized railway companies and only marginally influenced by public decision-makers. At the beginning of the 1920s, the political control of transport policy development increased. But the patterns of public actions of the players involved were predominantly automotive oriented. No *direct* influence was exercised on supporting public transport or on controlling the regional mobility patterns. However, it was precisely the non-action of actors that had a considerable impact on the electric railways and the mobility patterns of the metropolis and should therefore not be neglected. Section 6.3.1 explains the reluctance of the political–administrative actors to act with sociocultural factors.

The following section focuses on the interactions between the (car-oriented) transport policy and the automobile as a mode of transport on the one hand and mobility patterns on the other hand (Section 3.3). Before doing so, however, it outlines the transport policies with which the actors responded to the mobility problems of the 1920s (Section 3.1) and thereafter (Section 3.2). Whether the construction of a freeway network has solved the mobility problems of the region and fulfilled the hopes of its proponents is briefly discussed at the end of this chapter (Section 3.4).

3.1 Transport problems and transport policy in the 1920s

Although the coincidence of the innumerable trains of the LARY and the PE during the rush hour had already led to considerable congestion problems in 1910, with the increased emergence of the automobile in the CBD, "traffic crises" developed (Wachs 1992: 2). After the end of World War I, the number of cars in Los Angeles County quadrupled in about five years. Traffic censuses showed that in 1923, during a 12-hour period of a typical weekday, almost half of the trips to the CBD were made by car (cf. Bottles 1987: 204). In Chicago, by contrast, only 33% of people (1926) used the car to get to the CBD, with both business districts about the same size (Wachs 1984: 304; Bottles 1987: 93).

When in the summer of 1919 the workers of LARY and PE went on strike for higher wages, causing traffic chaos in the city center for several weeks, this enhanced not only the resentment against the railways but also the desire to own a car. The public now increased the pressure on the political actors to find solutions to the traffic problems. Businesspeople in particular feared that as accessibility to the center deteriorated, turnover and property prices could fall (Bottles 1987: 61 ff).

3.1.1 The fight against a nationwide parking ban

In a report requested by the Transport Committee and the Railroad Commission, transport engineers confirmed that the phenomenal surge in cars in the city center made it virtually impossible to operate trams. They made a strong recommendation to the city government to limit the number of cars in the center by imposing strict parking bans. In fact, the narrow streets had vehicles parked on both sides, leaving often only a small corridor for the tram and cars, with hardly any traffic flow. The Transport Committee endorsed the engineers' recommendations and called on the City Council to introduce *parking bans* in the CBD. This step was also intended to accommodate the LARY and avoid fare increases (Wachs 1984: 305). At the beginning of February 1920, the city council — after some concessions — issued a parking ban in the CBD from 11 a.m. to

6:15 p.m., which was to take effect two months later (Bottles 1987: 80). In view of the many cars in the city center, this was an instrument that intervened considerably in the everyday life of the citizens and was therefore very controversial in public opinion. This was not least because the ban marked the beginning of public regulation of road traffic (Brodsly 1981: 85).

The public debate on the parking ban, which will be briefly outlined in the following, not only revealed the various transport policy interests in the metropolis but also showed in an exemplary manner that there was no longer any successful policy to be made *against* the car. Influential players — from the automobile club, the industry and trade association to many businesspeople in the CBD — initially supported the initiative. They believed that this scheme could alleviate traffic problems. Studies showed that most cars parked in the center belonged to commuters and were not parked in the city for shopping. It was also thought that this would make delivery traffic in the center more efficient. Finally, they hoped that the parking ban would encourage many property owners to demolish their old buildings and build car parks (Bottles 1987: 66).[1]

However, many actors categorically rejected this initiative from the outset. Led by the Association of Car Dealers, they saw it as a law "that (...) would make the Street Railway Companies the sole beneficiaries" (Bottles 1987: 68). Many citizens and businessmen had also already lost confidence in the overcrowded and uncomfortable electric railways and regarded any regulation that could increase their revenues as negative. "Los Angeles Street Car Service Is Intolerable" was a headline in the liberal newspaper *Record*. It was reported that "condemnation of the traction companies of Los Angeles was never so general and bitter as it is today" (quoted in *ibid.*: 71).[2] The car, on the other hand, was increasingly seen as an alternative which could make the individual independent of these

---

[1] In fact, the world's largest car park opened its doors in the CBD at the end of 1925, offering space for 1000 vehicles on nine levels. Another "Sky scraper Garage" with 13 floors was opened in 1928 (Nelson 1983: 277 and 279).

[2] In those days, the Los Angeles Record reported in a series on the state of the railways and traffic in the city. Only once in the whole series were car owners mentioned as a factor for the poor traffic conditions. The Record as well as the Los Angeles Times rejected the parking prohibition. Also, "public opinion" was predominantly against such a regulation, which unilaterally favored the railways. It was believed that this regulation would lead to a loss of revenue for businesses in the CBD. It was also assumed that the already

conditions. According to the record, "Twenty years ago influential people rode in street cars (...) Today they ride in automobiles" (*ibid.*: 74). The car had developed into a symbol of flexibility and technological progress and enjoyed great popularity among the population.

When the parking ban finally became law on 10 April, the police issued almost 1,000 warnings — albeit free of charge — against incorrect parking and unauthorized left turns on the first day alone. After a few days, however, they wanted to impose administrative fines of between five and ten dollars. The scheme proved to be a complete success in terms of traffic flow. Trams, which previously lost 40–60 minutes per trip, were running on schedule for the first time in years. Vehicles passing through the center also saved 10–15 minutes during rush hours, and the number of accidents in the CBD fell drastically (Bottles 1987: 82 f).

Despite this positive balance, the already vehement opposition to the law that came into force increased. Retailers and the *Business Men's Cooperative Association* complained of a 25–35% loss of sales and wanted to take legal action against the parking ban. Only shops on the outskirts of downtown recorded their best sales in years, as many motorists parked their cars on the edge of the no-parking zone and shopped in the nearby department stores (Bottles 1987: 83 ff).

The protest grew stronger. Four days after it came into force, a motorcade drove demonstratively through the city center. A representative of the car sellers' association argued, "The day when the automobile was a 'pleasure vehicle' (...) is long since past. The motor car is just as much a necessity to business as the streetcar" (quoted from: Bottles 1987: 86 f). *The Examiner* commented the following:

> The people of Los Angeles want traffic and business expedited and not killed. The anti-parking ordinance is being condemned by merchants on the ground that it is ruining trade. If it is, the ordinance should be promptly amended (quoted in: Bottles 1987: 85).

Despite protests from the *State Railroad Commission* and the Transportation Committee, the City Council gave in to public opposition 19 days after the law came into force and replaced the original parking ban with a much more car-friendly regulation. Previously, cars were only

---

completely overcrowded railways would not be able to cope with the additional transport burden (cf. the detailed discussion in: Bottles 1987: 66–82).

allowed to stop for 2 minutes in the CBD, but now they could park for three quarters of an hour between 10 a.m. and 4 p.m. However, the strict parking ban between 16:00 and 18:15 was maintained. This decision was a clear defeat for the railways, which now had to push through congested roads again. But the victory of the motorists was also only symbolic, in the sense that it signaled to political actors that a transport policy cannot be made against the car. The victory did not change the fact that the traffic situation in the center continued to deteriorate for motorists.

After that, traffic regulations were introduced to improve the flow of traffic. Apart from the introduction of additional traffic rules (prohibition of left turns, use of traffic lights, one-way streets), pedestrians were also to be separated from road traffic by tunnels and bridges — according to an idea of the automobile club (Bottles 1987: 99 f). However, all these measures brought only short-term and marginal improvements to the traffic situation in the center. The population boom and the exorbitant increase in car density in Los Angeles called for more far-reaching solutions.

3.1.2 The road transport plan

There was subsequently a broad consensus among transport policy actors that traffic problems could be solved by substantially enlarging the road network and widening existing roads. Indeed, Los Angeles had a comparatively small road network in the city center.[3] However, the widening and new construction of roads required a complete *reorientation of the political action patterns that had been common until then*. Earlier, transport policy action had been extremely passive, reactive, and incremental. In contrast, the expansion of the road network required political commitment and at least a coordination of construction projects. However, coordinated construction of roads was not common. Until the 1920s — and Los Angeles was no exception — there was a complete lack of planning in the sense of a targeted and methodical systematic approach to anticipate the consequences of action. It was not until 1920 that the city of L.A. set up a *planning commission*, which was followed in 1923 by the

---

[3] Whereas in 1924 only 21.5% of the CBD area in Los Angeles was made up of streets, this figure was 44% in Washington, 34.5% in San Francisco, and 29% in Chicago (cf. Bottles 1987: 99).

first regional planning authority of the US at the county level (Wachs 1984: 306).

Until then, construction works were carried out individually and practically without reference to the capacity and structure of the general road network. The construction of a road had to be initiated by a majority of the affected landowners by means of a petition to the city council. The latter then commissioned the "Engineering Department", where the petitions were examined by civil engineers on a "case-by-case basis". If approved, a construction company was commissioned, and the adjacent property owners were taxed for the entire construction costs (cf. *ibid.*: 306 f; Folgelson 1967: 247 ff; Bottles 1987: 96). Although this procedure kept the city's expenditure low, it prevented coordinated traffic planning.

While the city council generally agreed to the new construction and widening of streets after the parking ban dispute, it hesitated to take concrete steps despite considerable pressure from problems. It argued that sound planning could save enormous sums of money (cf. Bottles 1987: 105). This hesitant attitude was certainly also an expression of a lack of action routines, since political actors in the transport sector had previously tended to be passive regulators, whereas now active and systematic action programs were required.

Significantly, the Road Transport Plan was driven forward by private initiatives. Frustrated by the City Council's continued delay in finding solutions to the pressing problems, the independent Traffic Commission[4] appointed 23 men to constitute the Major Highway Committee. Each of them donated $1000 to finance and design a comprehensive road traffic plan (Wachs 1984: 307). The committee hired the well-known city planners, Frederick Law Olmsted Jr., Harland Bartholomew, and Charles H. Cheney, who, from the many previous proposals and their own considerations, drew up the influential *Major Traffic Street Plan for Los Angeles* and presented it to the citizens in May 1924 (Nelson 1983: 277).

The report underlined once again with figures how urgent the problem of congestion on the CBD roads had become. Surprisingly, as early as

---

[4]The Traffic Commission was a voluntary association of business and civic organizations, utilities, newspapers, and city councils, which saw itself as a forum to discuss the various interests of the city. The aim was to help the city's decision-makers solve L.A.'s traffic problems. The Commission, founded in 1922, acted as a link between the public and the City Council, acting in the public interest. Elected decision-makers were not allowed to participate (cf. Bottles 1987: 101 f and 274, note 16).

1924, during 11 hours of a typical working day, almost the same number of people arrived in the center by car as by train. Car traffic had quadrupled in many of the county's thoroughfares within five years (Bottles 1987: 107). The planners' recommendations resulted in a *rationalization of traffic movement.* They considered congestion to be a consequence of "unscientific" road width, design, and effective use of road space. The road network and its capacity had to be designed in such a way that they resulted in "a balanced scheme for handling a tremendous traffic flow" (quoted from: Brodsly 1981: 85).

Consequently, the streets were differentiated according to their function into main thoroughfares, parkways or boulevards, and side streets. For the city to be able to continuously adapt to the respective developments, it would make no sense to fix and implement a fixed plan for all street types in advance. Rather, the *solution to L.A.'s congestion problems* lay in the development "of an orderly and well-balanced system of thoroughfares of such width and arrangements as will facilitate direct and uninterrupted movement from centre to centre and incidentally facilitate direct and uninterrupted movement within centres" (quoted after: *ibid.*: 87). The planners logically recommended a dense network of large six- or eight-lane thoroughfares with two parking lanes.

The report, which proposed a total of some 200 individual projects, attached great importance to the spatial separation of different types of traffic, as their mixing was considered to be a cause of congestion: Cars should not mix with the slow delivery traffic and trams should be separated from other traffic, where feasible, by elevated or underground railways. In contrast to many future road transport plans, this one also mentioned the importance of pedestrian traffic. For example, a narrow pavement for a main road was rejected as impractical (Brodsly 1981: 87).

The planners made it clear that, even by widening and broadening, the road network can reach its capacity limits with increasing traffic density, which meant that the problem of road congestion "(would) be just as bad as with smaller limit capacity (...) The street car, owing to its economy of space and low cost of operation per passenger", the planners said, "must take precedence over other forms of vehicles in the congested area whenever the traffic capacity of the arteries approaches its limits" (Bottles 1987: 110). In addition to the construction of underground railways in the CBD, they also considered the decentralization of the retail trade to be indispensable as a supplement to solving the traffic problems.

The planners recommended a dense network of roads and their proposals thus supported the development toward a dispersed, multi-central urban form. Condensation and concentration were seen as one cause of the traffic problems.

The first *scientific* road transport plan was generally seen as a means to solve transport problems. The city council unanimously agreed to submit the plan and a $5 million bond to the citizens for their vote. The plan and the first bond for construction were overwhelmingly approved by the voters (5:1 and 3:1, respectively). In the final annual report of the Transport Commission, the President wrote that citizens understood that congestion was the most serious problem facing the city and "that millions of dollars must be spent to convert our streets into a balanced roadbed capable of handling the enormous and mounting traffic of this modern era" (Bottles 1987: 113).

The city council's strategy of making only small sacrifices for the citizens and interest groups was successful. As early as 1926, a temporary levying of property taxes for the construction of roads by the citizens was confirmed. The rejection of all other measures put up for election at that time indicates a high level of popular support for road construction. In the following years, further laws were passed to facilitate the implementation of the plan (cf. Bottles 1987: 126 f).

But before the great depression began in the late 1920s, only a few projects had been implemented (Brodsly 1981: 89). Since hardly any financial resources were available, only a few construction projects were completed during the long years of the depression and during the Second World War. The few increases in road capacity could not cope with the continuing population and especially the car boom of the 1920s, so the traffic problems of the city and especially of the CBD tended to worsen. In 1930, the Transport Commission reported, "traffic conditions, particularly in the downtown area, are becoming chaotic, in fact so much so that business interests and general public are complaining bitterly in many instances" (quoted from: Bottles 1987: 118).

3.2 Freeways as a solution to traffic problems

The Road Transport Plan was the transport policy strategy to tackle the transport problems of the 1920s and beyond. Although attempts to finance it failed, it was the pattern on which the subsequent planning and political action was based. In the period that followed, the intention was to solve the metropolis's traffic problems once and for all by means of freeways. While the Depression and the Second World War slowed down population and automobile growth, in the post-war period, the population exploded from 3.2 million (1940) to 7.5 million (1960), as did automobile ownership. Increasing suburbanization produced exorbitant (car) traffic flows which flooded the already congested streets of the metropolis and in turn escalated into a "traffic crisis" (Wachs 1992: 2). The unsolved traffic problems in the center accelerated the decentralization of trade and industry. Nonetheless, the same traffic problems as in the CBD seemed to recur in the booming business districts of the surrounding area in the 1930s (Bottles 1987: 212).

In the surrounding areas, too, it became clear that the extension of roads alone was not an adequate means of solving congestion problems. On the contrary, as early as the 1930s, planners, public decision-makers, and other transport policy actors favored the construction of *expressways* or *freeways*, of roads — certainly oriented toward the German *Autobahn* (Brodsly 1981: 97) — with considerably more capacity. Once again, private actors took the initiative and pointed out the urgency of finding solutions to transport problems. In all studies, the automobile was considered the most important mode of transport. By the end of the 1930s, automobile ownership had already exceeded the one million mark in Los Angeles County. A study published by the Automobile Club in 1937 stated that 80% of all driving was done by car[5] (Brodsly 1981: 98).

3.2.1 Freeway construction as a solution to the problem

In its study, the Automobile Club recommended the construction of a new type of road, later called "freeway", to solve mobility problems. Entirely in the style of the 1925 Road Traffic Plan, a *scientific* analysis of the problem preceded the recommendations: It was due to the *mixing of traffic from different directions* at intersections. According to this definition, the construction of more roads could not solve the problem, as it would simultaneously create more intersections, which would only make the problem

---

[5]Characteristically, the study was limited to the analysis of motorized modes of transport. Therefore, only transport services (journey lengths as an indicator) were compared. 80% of passenger miles were covered by cars, 17% by trains, and 3% by buses.

worse. The solution was a network of four- to six-lane roads where level crossings could be avoided by means of access and exit ramps. Lower-level roads were to be passed by bridges or subways so as not to disturb the flow of traffic on the motorway (Automobile Club 1937: 279). For the Automobile Club, mobility problems were thus merely an expression of *traffic engineering errors*, which in turn could be technically corrected.

What was striking about the proposed route was that it was strongly based on the lines of the railways. For example, freeways were to be built along all the steam railway lines of the *Southern Pacific* and the *Santa Fe Railroads*. The freeway routes also followed the lines of the regional railways in many places. The automobile club had intensified the debate with its study, which resulted in the demand for a general transport plan commissioned by the city itself. In 1939 the members of the *Transportation Engineering Board* appointed by the Mayor of L.A. presented their study, which was to meet the demands to all relevant interest groups in the city (Bottles 1987: 218 f).

The study "A Transit Program for the Los Angeles Metropolitan Area" was strongly based on the Traffic Survey conducted by the automobile club. Like the latter, the *Transportation Engineering Board* saw the city's traffic problem in the loss of time at intersections, which it estimated at around 30% of the total travel time (1939: 2). The routes of the recommended "express-highway"[6] system were also based on the study of the automobile club. In contrast, however, this plan was more detailed, both in terms of routes and funding and in terms of the political implementation strategy. It was argued that the adoption of the whole plan was less important than the immediate construction of priority routes. Thus, the report proposed as a first recommendation the construction of a *Hollywood car park*,

> because, in its opinion, there would be combined the greatest number of advantages from the standpoint of the public. This route would connect two of the district's most important centers of trade and population and would be exclusively a City of Los Angeles project thereby making

---

[6]The report uses the term "express-highway", but stresses that "express-highway", "motorway", "freeway", and "limited way" are used synonymously by many authorities, although the design standards vary somewhat (Transportation Engineering Board 1939: 4). In the following years, however, the term "freeway" became widely accepted.

for speed of decision, construction and use (Transportation Engineering Board 1939: 15).

In contrast to the Automobile Club study, this plan recognized the spatial advantages of buses and trains compared to cars and also drew attention to the social dimension of automobility:

(S)o long as the number of automobiles is only half the number of adult persons in the district, the use of the family automobile by anyone of the family leave the remaining members without transportation. For this and other reasons, mass transportation are vital to the existence of large communities (*ibid.*: 4).

The planners accordingly recommended the controlled use of buses on the motorways (*ibid.*: 7). While referring to the transport advantages of railways, they anticipated that "(t)he high cost of the most approved rail arrangements tends to defer into the indefinite future the time when they can be financed" (*ibid.*: 14).

3.2.2 Public acceptance for freeway construction

The plan received approval from all sides. Many citizen groups supported the proposed extensive freeway network because they believed it would also protect their interests. The freeways were seen as a solution to the urban diseases of dense, congested cities, allowing vehicles to flow freely between the now numerous subcenters. The CBD's stakeholders were also satisfied with the plan, particularly as it provided for a dense radial network in the center of the city, which seemed to increase its accessibility for customers who had been turning away since the 1920s. Although in the past they favored an elevated and underground railway system for the center, businesspeople in the CBD — with the declining chances of its financing — slowly began to come to terms with the freeway proposals. Public transport could, they argued, also be improved by using buses in the center. For businesspeople in the CBD, this plan thus seemed to be "the easiest and quickest way to renew the city's centralised spatial structure" (Bottles 1987: 223; cf. *ibid.*: 221 ff; cf. also below: Section 6.2).

Although agreement had not yet been reached on the financing of the freeways, public consensus on this seemed much easier to achieve.

For example, the *Los Angeles Department of City Planning* celebrated the *Transportation Engineering Board's* freeway plan two years later as a radical development exclusively for automobiles and as an "innovative" response that had become necessary for Los Angeles' "innovative life-style" (Brodsly 1981: 105). Although the *County Regional Planning Commission* also accepted the plan in 1943, there was virtually no construction activity during the Second World War (1942–1945). The planning of freeways was — as was the case with roads before — much easier than their financing.

After all that, the *Arroyo Seco* (later Pasadena Freeway), the first 6.8-mile-long section of a freeway, was opened at the end of 1940. Half of the $5 million construction cost was borne by the *State of California* and the other half by federal and city funds as well as by L.A. and Pasadena's petroleum tax revenues. Despite the universally accepted freeway plans, only 11 miles of freeway were opened in Los Angeles at the end of the Second World War, followed by a further 4.3 miles by 1950 (Brodsly 1981: 112).

3.2.3 The financing of the freeways

The financing of the freeway projects was made possible by two new regulations. First, after numerous hearings and studies, the State of California passed the *Collier-Burns Highway Act* in 1947, which aimed to ensure the financing of highways by raising mineral oil taxes and car registration fees. Through this Act, the *state* was involved in the construction of freeways by being responsible for financing rights of way, planning, construction, and maintenance of urban highway routes (Brodsly 1981: 115). The *Collier-Burns Act* now ensured funding for years to come. California had an annual budget of some $76 million for this purpose, and the *Los Angeles City Planning Commission* estimated that the city would receive a total of $300 million over the years for the construction of the freeway system (Bottles 1987: 233). Between 1950 and 1955, when large parts of the Hollywood–Santa Ana and San Bernardino Freeways were opened, the network of freeways in the metropolis grew more than fourfold.

The *Federal Aid Highway Act* of 1956 brought a second, even more important construction thrust, which, with the construction of a "National System of Interstate and Defense Highways", represented the largest public project ever undertaken in the USA (Weiner 1987: 14). With the planned construction of 41,000 miles of Interstate and Defense Highways, it aimed primarily at connecting larger cities. Between 1957 and 1969, $24 billion was to be spent on the construction, 90% of which was to be covered by federal funds. Weiner stressed that urban areas were the main beneficiaries (1987: 15). In Los Angeles, for example, the San Diego, Golden State, Santa Monica, San Bernardino, Foothill, and San Gabriel River Freeways were built under this program. In this context, it should be noted that at that time and for many years afterward the federal government in Washington did not provide any funding for public transport, unlike for roads.

A new source of funding opened in 1959 with the *California Freeway and Expressway System Act*, which provided for a dense freeway network for the Los Angeles region. When completed, no urban area was to be more than four miles from a freeway access road. The *State Division of Highways District VII*, which was responsible for the counties of Los Angeles and Ventura, was then to build a network of freeways totaling 1557 miles, which was to be completed in the 1980s (Brodsly 1981: 119). By the end of 1994, over 600 miles of these had been built.

Figure 3.1 shows that the construction of freeways reached its peak between 1965 and 1970. The increase in California's petroleum tax

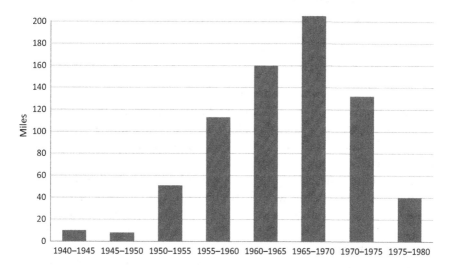

Figure 3.1:   Length of the freeway network in five-year periods.

Source: Own illustration based on Brodsly (1981: 126).

by 1.5 cents per gallon in 1956 and by a further cent in 1963 to a total of 7 cents per gallon and a further 4 cents levied by the federal government ensured this development.

Since the late 1960s, however, public approval for the construction of new freeways had been declining. In addition, the construction funds, which were so abundant, were slowly running out. The 7 cents per gallon levied by the *state* since 1963 melted down to 2 cents in real terms in 1979 (Brodsly 1981: 120). Attempts to levy further taxes on the construction of freeways did not find support among the population. Brodsly is right when he wrote that "(t)he age of new freeway construction has, for the most part, ended" (1981: 120). On the contrary, since the end of the 1960s, with the partial deterioration of the traffic situation and air pollution, doubts were raised about the solution of traffic problems by freeways at all. Whether these doubts were justified is briefly discussed in Section 3.4.

Since that time, public transport in Los Angeles has again been the subject of increased discussion. It is true that express buses have long been using the metropolis's numerous freeways. But amidst the dense automobile traffic, "they produced all the inconveniences of public transit without affording any of its benefits" (Brodsly 1981: 158). Since the late 1960s, attempts have been made to make public transport more attractive again. Among other things, the first eleven-mile express bus lane was built in 1974 on the San Bernardino Freeway (see also Chapter 4).

3.3 Mobility patterns

The following section examines how the car, as the dominant mode of transport in the metropolis, affected mobility patterns. As the car has been the dominant means of transport since the 1920s, the analysis also covers this period. With the aim of identifying mobility patterns, the relevant development steps are outlined. In the following, the mobility structure changing during the car phase and especially during the construction of the freeway will be dealt with first. The focus here is particularly on the spatial arrangement of the activity locations of living and working. The analysis of the mobility behavior of the citizens should then provide information about the development of the modal split, journey length, and journey times. Finally, Section 3.3.3 outlines the mobility opportunities of social groups in a car-oriented society.

3.3.1 Mobility structure

3.3.1.1 *Reinforced settlement*

The population boom of the 1920s and after the Second World War caused the metropolis to grow explosively once again. While the central areas grew at a comparatively low rate in the following decades, the growth rates in the outskirts were extraordinarily high. For example, the population of Orange County (southeast) alone tripled between 1920 and 1930 to over half a million people (cf. Figure 3.2).

As the Angelenos were dependent on the mobility of the railways, their homes were always relatively close to the railway lines. Freed from such constraints by the car, citizens could henceforth settle in the vast free and now accessible areas between the railway lines. Characteristically, a comparison of road maps from 1919 and 1930 shows only "modest 'filling in' of streets in sections where the trolley lines were seldom more than two miles apart. On the other hand, the automobile's triumph exerted a dramatic effect on the remote areas which were not so well served by the

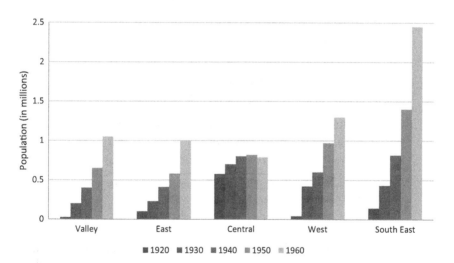

Figure 3.2: Population growth in various areas of the metropolis of Los Angeles (1920–1960).

Note: The area of the metropolis is limited in this graphic to the counties of Los Angeles and Orange. The composition of the different counties is the same as shown in Table 3.1.

Source: Own illustration based on Pegrum (1968: 565).

Table 3.1:   The Central Business Districts (CBD) in the five
largest metropolitan areas in the US (1980).

Metropolis	Total number of jobs in the metropolis (thousands)	Share of jobs in the CBD (percent)
New York	6,628	10
Los Angeles	*4,366*	*4.2*
Chicago	2,990	9.7
Philadelphia	1,689	12.2
San Francisco	1,537	14.7

Source: Gordon (1991: 12).

Table 3.2:   Traffic volume in four business
centers in 1941 (within a 16-hour period).

Mode of transport	Number of journeys	
	CBD	Long Beach
Car	482,000 (54.9%)	127,000 (63.3%)
Railway	275,060 (31.4%)	27,000 (13.5%)
on foot	120,000 (13.7%)	46,000 (23%)
Total	877,000 (100%)	200,000 (100%)
	Pomona	**Westwood**
Car	36,000 (82.5%)	47,360 (87.1%)
Railway	670 (1.5%)	4,000 (7.4%)
on foot	7,000 (16%)	3,000 (5.5%)
Total	43,670 (100%)	54,360 (100%)

Source: Regional Planning Commission, Business
Districts 1942, quoted below: Bottles (1987: 205).

trolleys" (Foster 1976: 477). The best example of this is the San Fernando
Valley, where between 1920 and 1930 the population increased more than
fivefold; many areas, which were often far more than two miles apart
from the regional railways, were developed and thus dependent on the car
as a means of transport (cf. *ibid.*). Table 3.2 illustrates the considerable
acceleration in population growth in the suburbs in the following
decades.

Growth in the central areas remained modest throughout, and even saw a slight decline in population between 1950 and 1960. The surrounding areas, on the other hand, had to cope with considerable population growth.

A characteristic feature of the population boom in Los Angeles was the consistently high proportion of single-family homes. In 1930, Los Angeles had the highest percentage of single-family homes (93.9%) compared to cities of comparable size. While Philadelphia was the only city with a similar ratio of 91.6%, this proportion was much lower in Detroit (79.7%) and Chicago (52%) (Fogelson 1967: 146). The high proportion of single-family homes was still typical of the *style of low-density living in* Southern California. The open space between Los Angeles and its small suburbs disappeared more and more during settlement, resulting in a uniform suburban *sprawl* with no discernible beginning or end.

Since the 1920s, new residential areas have been built that were only geared towards the car as a means of transport.[7] Like the famous *Palos Verdes Estates*, areas were designed as "garden suburbs", separating residential streets from main roads and designating large areas as parks. Commercial areas, on the other hand, were provided for only a few people, while industrial settlements were not planned at all (Fogelson 1967: 157 ff). In fact, most of the newly developed settlement areas at that time were exclusively residential areas which were deliberately separated from commercial and industrial activities in order to make them more attractive for potential buyers (*ibid.*: 156 f).

The construction of freeways in the 1950s, 1960s, and 1970s greatly accelerated the dispersal of the population into the region. Rapid population growth in the remote areas of the San Fernando Valley, San Gabriel Valley, South Bay, and Orange County coincided remarkably with the construction of the freeway. The completion of the Santa Ana and San Diego Freeways effectively linked Orange County with Los Angeles County "to form a single megalopolis" (Brodsly 1981: 12). Along the freeway routes, small residential towns formed almost everywhere; Irvine and Westlake are good examples of this. The freeways were mostly built through sparsely populated, cheap areas for cost reasons. Since land prices were much lower there, broad sections of the population were able to fulfill their dream of "houses in the country":

---

[7] Whereas a few years earlier real estate advertisements mentioned the proximity of railway lines, the main focus was now on the accessibility of areas by road (Foster 1976: 478).

Here rurality proves to be a supreme middle-class value, as many workers willingly assume daily drives lasting one or two hours in order to live in rugged and picturesque settings of mountain sheltered valley locations, free from smog and blighted landscapes (Brodsly 1981: 12 f).

The population growth rates, and the settlement patterns described have continued to this day. Between 1950 and 1990, the metropolis grew from 4.8 million to over 14 million inhabitants, with Los Angeles showing very homogeneous and very low settlement densities compared to other metropolises in the US (cf. above: Figures 3.1 and 3.2).

3.3.1.2 *Decentralization of trade and industry*

The automobile had thus exorbitantly intensified the trend toward urban sprawl created by urban and regional railways. Until the end of the 1920s, jobs in trade and industry remained concentrated in the CBD and in the area around the port. The enormous demand for mobility created by the one-sided use of land — working here, living there — was increasingly covered by private cars, starting in the 1920s, making the already congested CBD — as described earlier — increasingly inaccessible.

However, the car was already so widespread that department shop owners considered the inaccessibility of their shops to this mode of transport a competitive disadvantage. John Bullock, for example, the owner of a large department shop chain, was one of the first in his industry to decide to move a large part of his business out of the CBD in 1928. The new location, a few miles west of the center, had a large free car park and quickly became popular with the population. This success convinced many other businessmen, so within the next five years, 88% of all new retail outlets were opened in the suburbs. While 75% of Los Angeles County's retail sales were still recorded in the CBD in 1929, this share dropped to 54% ten years later. By 1956, the figure had fallen to 23% (Bottles 1987: 196). Nelson and Clark calculated that — with the metropolitan area (SMSA) as a reference — only 3% of retail sales made in 1963 were in the CBD. The centers of other cities (15.5% in Manhattan, New York; 7.6% in Chicago) showed much higher sales figures by comparison (1976: 27).

This declining importance of the CBD was also reflected in the distribution of jobs. At 43%, the financial sector still had a fairly high concentration of jobs in central L.A. In contrast, the *consumer-oriented* services provided by banks and insurance companies followed the settlement pattern of the population and explained the distribution of the other half of jobs across the region (cf. Nelson/Clark 1976: 44–46). Similarly, jobs in the wholesale trade were increasingly located near the center. Nelson and Clark assumed that the freeway system, which was laid out radially toward the center, was the cause of this sector, which was heavily dependent on accessibility. In contrast, the distribution of jobs in industry was quite evenly spread throughout the metropolis. Trucks were largely responsible for this. Its maturity coincided with the expansion of the industrial sector in the 1920s. While there were only 10,100 trucks registered in the US in 1910, this number rose to 1.1 million ten years later and tripled again to 3.6 million by 1930 (Motor Vehicles Manufacturer Association 1977).

Agriculture and tourism were still the dominant economic sectors before the First World War. Compared to other cities, Los Angeles therefore still had a comparatively meager industrial production. From 1909 to 1929, however, the value of goods produced rose from $68.6 million to over $1.3 billion (Fogelson 1967: 124; 129). Thanks to the oil-producing and processing industry, car and tire production companies, and the film industry, the metropolis developed into the largest industrial center in the West.[8] The lorry gave above all these space-intensive companies the opportunity to locate outside the center, especially as land prices near the center rose considerably. Ground-level production was in any case more efficient than a multi-level structure in a small area (Bottles 1987: 198). This made it possible to locate companies on cheaper land at the periphery, because in addition to the truck as the metropolis's freight carrier, the car also gave workers the opportunity to get to the factories far from the railway lines. In 1938, for example, 71.7% of commuters working outside the center already used their own car, while only 20% used public transport and the rest came to work by bicycle or on foot (Transportation Engineering Board, cited in: *ibid.*: 198 f).

---

[8] Henry Ford moved a business opened near the city center in 1917 to Long Beach in the south of the metropolis ten years later. Other car manufacturers followed, giving the metropolitan area the largest volume of car production after Detroit (Nelson 1983: 185). The Good Year Tire and Rubber Company, Tirestone, and Goodrich also opened large complexes to produce tires there (cf. also oil and film companies in Los Angeles, *ibid.*: 180 ff).

Table 3.3:   Modal split of trips to work in the Los
Angeles metropolitan area.

Mode of transport	Year		
	1970	1980	1990
Car	85.9%	85.5%	85.6%
PUBLIC TRANSPORT	5.5%	7.0%	6.5%
on foot, bicycle	8.6%	7.5%	7.9%

Source: Census of Population: Journey to Work, Department
of Commerce, various years.

The construction of the freeway network increased the dispersion of trade and industry since the 1950s. Most locations were now located near freeways. New as yet undeveloped areas offered themselves as locations with the completion of the project. Disneyland, for example, decided on the location directly next to the Santa Ana Freeway after an extensive study (Brodsly 1981: 18–22).[9] Despite the construction of a wide ring of freeways — which simultaneously defined and symbolized the center (Brodsly 1981: 23) — only around 4.2% of jobs were still located there in 1980 (as the largest center). This was very little compared to other metropolises (cf. below: Table 3.3).

Of particular importance for transport policy were the spatial patterns of employment centers throughout the metropolis. Guiliano and Small counted 31 subcenters of the metropolitan area, in addition to the CBD, which, at 1.49 million, nevertheless accounted for less than a third of total employment (1990: 11). In addition to the CBD as the largest place of employment, the four largest main employment centers were located very close to the center, which was characterized by freeway radii. In general, most of the 32 subcenters of the metropolis were in the immediate vicinity of freeways (cf. Guiliano/Small 1990: 7). Characteristically, however, more than two-third of the region's jobs were not located in subcenters but were scattered beyond the metropolis. "Pure sprawl around a large

[9]Of course, proximity to a freeway was not the only criterion that determined whether a company was economically viable. Rather, the example of the Harbor Freeway leading through the ghetto, which was mainly inhabited by poor black people, made it clear that a freeway did not have to contribute to economic prosperity (cf. below: Section 3.3.3). The construction of a freeway had the greatest effect on increasing prosperity when the area was already well developed. In this respect, the construction of a freeway had a strengthening effect on economic development (cf. also Brodsly 1981: 22 f).

corridor-shaped core, more than subcentering, best describes the location of the majority of the region's jobs" (Guiliano/Small 1990: 10).

This distribution pattern of jobs, together with the dispersion of the Angelenos' homes, meant a considerable thinning of destination and source traffic. Due to the very low passenger volumes to be expected for the respective connections between living and working, public transport could only operate with considerable subsidies. In particular, means of transport with high transport capacities (underground, rapid transit) appeared rather unsuitable under these conditions (cf. below: Section 4.2.1).

The region was also characterized by a strong spatial imbalance between residential areas and places of work. For example, the ratio of jobs to households was 1.47:1 in the area with the most jobs in downtown L.A. (Guiliano/Small 1991: 12), while in residential areas such as North Los Angeles and San Bernardino Desert, it was 0.71:1 and 0.56:1, respectively (SCAG 1989a: VI-8). Workers who had their jobs in downtown and other subcenters commuted the furthest and for the longest time on average (Guiliano/Small 1991: 15). This result was in line with the general expectation that large concentrations of jobs compared to the surrounding population attracted workers from a geographically large area. By contrast, areas with a low job-to-household ratio had shorter commuting distances and shorter commuting times. Companies could draw on a large pool of workers in these areas (*ibid.*: 17; see also below: Section 3.3.2).

3.3.2 Mobility behavior

The outline of the mobility structure essentially shows that, on the one hand, the tendency toward urban sprawl increased considerably since the 1920s and, on the other hand, two-third of the places of work, which has been previously concentrated mainly in the center, were now scattered throughout the metropolis. Only one-third of jobs were now to be found in the 32 subcenters. In addition, a spatial separation of residential and work activities had been observed. The mobility needs indicated by this have already been pointed out. The following section examines how these mobility structures affected the mobility behavior of citizens. In this context, the interrelationships with modes of transport must necessarily be discussed again.

In the 1920s, the car established itself as an indispensable means of transport. After the First World War, the growth rate of cars had already far exceeded the growth of the population. While in 1920 only one in six people in L.A. County owned a car, this figure rose by 1930 to more than one in three (2.75). In a comparative perspective, the motorization rate in L.A. County was much higher than in other cities (Chicago 1:8) or as the US average (1:5.3) by 1930.

Accordingly, the ratio of car ownership to households increased steadily after the Second Word War. While this ratio rose from 1.21 to 1.33 in the metropolis between 1960 and 1970, only ten years later households had an average of 1.7 cars (Cervero 1986: 21). Los Angeles was thus still one of the most densely car-equipped cities in the US.[10] In the Los Angeles area, the average household had 2.8 persons, so the car density in 1980 was around 1.6 inhabitants/car. In 1990, the car density further increased to 1.42 inhabitants/car. In Los Angeles County, the number of registered cars exceeded that of holders of driving licenses.

Traffic censuses in various business centers of the metropolis had already showed, at the end of the 1930s, that they were mainly reached by car. Only the CBD still had a comparatively high proportion of public transport, with a good 30% train journeys in 1941 (Table 3.2).

Surprisingly, the modal split in the CBD hardly changed forty years later, despite the construction of hundreds of freeway miles, many of which ran through the center. For example, traffic censuses from 1980 showed that the CBD still attracted most travel, with 690,000 trips, of which about 66% were made by car, 27% by bus, and 7% on foot (OECD 1988: 83). In 1984, only 5% more people visited the CBD than in 1955, with a 23% increase in the number of cars over the same period. However, whereas in 1939 the average number of people in a car was 1.51, this proportion fell to 1.46 in 1955 and to only 1.36 in 1984 (Woodhull 1991: 7 f). The considerable space required by cars meant that, despite the enormous expansion of road capacity, the flow of traffic hardly improved as a result.

After the discontinuation of rail-based public transport, the bus was the only motorized means of public transport since 1963 (until 1990). While the number of car commuters remained roughly the same between 1970 and 1990 at just under 86%, the share of public transport commuters had risen by only 1% since 1970 and later stood at 6.5% in the metropolis (1990). Around 8% walked or cycled to work (cf. Table 3.3).

---

[10] In Houston, the car density was even higher in 1980 at 1.9 cars per household, and in Denver and Dallas at 1.8 cars (cf. Cervero 1986: 21).

The following mobility behavior was typical for public transport users: They drove an average of only 12 miles to work (16.6 for all modes of transport; Commuter Transportation 'Services 1992: 7) but took considerably longer than drivers. On a national average, commuters on public transport took more than 42 minutes to get to work, which was twice as long as drivers who only needed 21 minutes (Pikarski 1987: 57).

Public transport commuters with the destination Central L.A. used the bus far more than proportionately, which indicated, among other things, a radial orientation of the bus lines. This was also confirmed by Burns and Harman in a 1968 study: "The increased availability of bus service nearer the city center means that mass transit use increases with increased proximity to downtown" (1968: 32). In contrast, workers traveling to West L.A., South L.A., and the San Fernando Valley used the car up to 90% of the time, with feet as a mode of transport sometimes, accounting for a larger proportion than buses (Nelson/Clark 1976: 56). Residents of areas outside L.A. County used significantly less public transport (7% for L.A.; Orange, Riverside San Bernardino, Ventura Counties between 1–2%) and, in contrast, traveled to work by car more often (Commuter Transportation Services 1992: 50).

In the 1950s and 1960s, the mobility radius of citizens initially increased considerably. Between 1953 and 1962, for example, the area accessible from the center within 30 minutes by car expanded from 261 to 705 sq. miles (Nelson/Clark 1976: 52). This increase of around 175% was the result of the completion of freeways which allowed higher speeds and thus enabled more distant destinations to be reached at the same time. However, the accessibility of the metropolis deteriorated again with the considerable increase in traffic volume in the 1960s, 1970s, and 1980s.

The greater accessibility of the metropolis changed — as described — the mobility structure of the population, which in turn was reflected in mobility behavior. The distance between home and work was considerably greater in Los Angeles than in other metropolises. Whereas in 1970 in Philadelphia around 20% and in New York around 15% of commuters traveled only five miles to their workplace, in Los Angeles this figure was only around 5% (Nelson/Clark 1976: 53). This trend has remained constant to date. The average distance between living and working in the metropolis in 1992 was 16.6 miles (Commuter Transportation Services 1992: 7). On a national average, Americans commute only 10 miles (one way) daily (Pikarski 1987: 60). 32% of commuters in Greater L.A. drive

Table 3.4: Average distance and travel time per capita for the easy way to work in the metropolis of Los Angeles 1992.

County	Distance	Travel time
Los Angeles	15.8 miles	37 minutes
Orange	14.4 miles	32 minutes
Ventura	17.7 miles	28 minutes
San Bernardino	20.5 miles	35 minutes
Riverside	20.9 miles	38 minutes

Source: Commuter Transportation Services (1992: 51).

more than 20 miles to work. As expected, workers in Riverside, San Bernardino, and Ventura Counties (rather residential areas with a low work-to-household ratio) commute further than those in the "job-rich" Los Angeles and Orange Counties (see Table 3.4).

On the other hand, the different lengths of the distances to work hardly affected the commuting time: While the average commuting time to work in Greater L.A. was 36 minutes, the longer commutes of workers residing in Riverside, San Bernardino, and Ventura Counties did not result in longer commuting times. However, citizens of the L.A. metropolitan area drove to work longer than the national average, which was only about 22 minutes by car (Pikarski 1987: 57). It is also striking that workers who used freeways on their daily commute to work traveled a further 24 miles on average and took 46 minutes longer than the national average (Commuter Transportation Services 1992: 9, 11).

3.3.3 The selective effect of automobility

The following section examines how the distance-intensive mobility structures with car-oriented mobility have affected the mobility opportunities of different socioeconomic groups. Car-oriented structures naturally also have a considerable influence on the mobility opportunities of other social groups, especially the elderly, the disabled, women, and children. For reasons of scope, the analysis is largely limited to socioeconomic differences. To answer this question, we will first briefly look at the unrest in Los Angeles in the 1960s, as this was the first time that the links between poverty and mobility were widely discussed in public. Then, the different mobility opportunities of the Angelenos will be worked out.

3.3.3.1 *Poverty and mobility: The riots of the 1960s*

The link between access to transport and unemployment — and poverty as a result — first came to public attention following the 1965 racial riots in Watts, South Central Los Angeles. Between August 11 and 15, 34 people died, over a thousand were injured, and nearly four thousand were arrested in bloody conflicts (Fogelson 1971: 191 f). The *McCone Commission* (named after its chairman, the well-known entrepreneur and former CIA director John A. McCone), established by Governor Brown just four days after the riots ended, produced its report "Violence in the City — An End or a Beginning?" (in the following quoted: McCone 1966), which caused a lot of criticism, among other things because of its partly superficial analysis of the causes of the riots (cf., e.g., Fogelson 1971: 191–216).

The Commission identified as one of the main causes of the unrest, in addition to unemployment and poor education, the inadequate and expensive transport system in Los Angeles, which limits the mobility of the citizens of South Central:

> This lack of adequate transportation handicaps them in seeking and holding jobs, attending schools, shopping, and in fulfilling other needs. It has had a major influence in creating a sense of isolation, with its resultant frustrations, among the residents of south-central Los Angeles, particular the Watts area. Moreover, the lack of adequate east-west or north-south service through Los Angeles hampers not only the residents of the area under consideration here but also of all the city (McCone 1966: 145).

Since many citizens in Los Angeles could not afford a car, they were dependent on bus transport. However, the Commission says that bus transport was totally inadequate and too expensive. In the Watts area, four different companies (including three public companies) operated regular bus services under concessions granting exclusive operating rights for certain areas. However, there was no coordination between the lines, which did not allow transfers between them, so a resident had to use different bus routes and pay the fare several times just to leave the area (see *ibid.*). Los Angeles was the only metropolis in the US not to subsidize its public transport system, so bus companies, faced with losses, cut their networks and raised fares.

In order to improve the public transport system, the Commission recommended the merging of all small bus companies into the Southern California Rapid Transit District (SCRTD), which would be subsidized from now on. It also proposed to extend the network, especially in the Watts area (McCone 1966: 145). The US Department of Housing and Urban Development provided $2.7 million for the district in May 1966, so the recommendations could be partially implemented (Fogelson 1971: 206). A newly established bus line was successful and attracted over 3000 passengers daily. It was estimated that the bus line linking Watts with the industrial areas to the west (including General Motors) helped some 1200 people find employment during the three years of the project (Altshuler 1979: 276). Many of those who came to work through the new link did of course buy a car from their first earnings. After the federal authorities stopped their subsidies in 1971, when the project ended, many of the established lines were abolished without any noticeable improvement in the situation for the mainly black inhabitants of South Central Los Angeles.

In this context, it is important not to give the impression that an inadequate public transport system is considered the cause of the unrest in Los Angeles. Nor can improvements in this respect solve the precarious economic situation of the residents of south-central L.A. Fogelson agreed when he wrote, "The south-central ghetto is isolated, but not for reasons as simple and reassuring as dreadful bus service" (1971: 207). The reasons certainly lay much deeper. The spatial isolation of blacks was largely a consequence of racial discrimination. In this respect, for Anglo-Americans, the regional trains and then the car were a welcome means of moving to the suburbs, allowing them to distance themselves from the dirt, high crime rate, and heterogeneity of the inner cities. The sparse public transport system was an expression not only of spatial but also of socioeconomic isolation of poorer sections of the population. Certainly, however, the spatial immobility of these strata, especially in Los Angeles' mobility structures, which were highly distant, further worsened their job opportunities.

3.3.3.2 *Selective mobility*

In a metropolis with a mobility structure that is highly distant, such as Los Angeles, there is a strong relationship between individual mobility behavior and access to a motorized mode of transport. Since the public bus system in the metropolis was unanimously considered to be very poor (cf., e.g., Steiner 1978: 52) and was only rudimentarily modernized and expanded at the end of the 1970s, the *mobility* of the individual depended heavily on the availability of a car.

Although the number of cars in Los Angeles County exceeded the number of license holders in the late 1980s (Wachs/Guiliano 1992b: 3) and in 1990 statistically only about 1.4 inhabitants shared a car, many citizens did not have access to it because the distribution of car ownership was very uneven. In 1970, for example, 15.1% of households had no car at all, and in 1990 over 11% were still not car owners (US Bureau of the Census 1974, 1990). Households with high per capita incomes, such as Calabasas District ($38,204 per capita/year), had two or more cars in about 77% of the households, while about 20% of households in the black neighborhoods of East Los Angeles ($6,612 per capita/year) or Florence-Graham ($5,407 per capita/year) had no car at all (*ibid.*).

Since there was a high positive correlation between income and car density, low-income households also made significantly fewer car trips[11]: Brodsly, referring to a 1967 traffic study, pointed out in an extreme example that households in L.A. with a per capita income of less than $3,000 per year per household[12] made an average of only 2.2 trips a day, while households earning more than $25,000 made more than 12 trips a day. The upper half of the income groups also made more than twice as many (10.2 trips) as the lower half of the income groups, with only 4.8 trips daily (Brodsly 1981: 26).

These figures were also confirmed in a national comparison (Altshuler 1979: 268 f). Altshuler pointed out that within the same income groups, "members of households that own automobiles make two or three times as many trips as the members of households without them (...) Automobile ownership increases with income, so part of the increase in trip making

---

[11] Unfortunately, no data were available on the mobility rate, i.e., the distances traveled per inhabitant and time unit. A vast majority of statistical surveys on mobility have in the past and to a large extent still today been limited to the analysis of motorized modes of transport. However, since Los Angeles is characterized by a mobility structure that is intensive in terms of distance, the frequency of car journeys is also meaningful.

[12] With reference to Census data in 1970, Altshuler defined poor (or low-income) households as those with incomes below $5,000 per year (Altshuler 1979: 266).

associated with the former is reasonably attributed to the latter" (Altshuler 1979: 267).

The level of income was also expected to correlate with the length of the trips made. While the total population covered 60.6 person miles on average, the lowest-income group covered only 17.8 person miles daily (*ibid.*: 269). In terms of income, both Los Angeles and New York showed the expected trend for higher-income families to travel further for work. However, all income groups in Los Angeles traveled further than those in New York on average (Nelson/Clark 1976: 53).

As expected, suburban residents had a higher average car rate compared to Angelenos living closer to the city center and used private transport more often. For example, low-income groups used public transport significantly more often than "higher income commuters living in sounder (suburban) housing (which) are less likely to utilize mass transit facilities than those occupying poorer quality (central) housing" (Burns/Harman 1968: 32). This was already apparent from the region's public transport network, which could practically only be spoken of in the denser areas of the center. In the sparsely populated suburbs, people were almost completely dependent on the automobile. Toward the CBD, however, the bus network became denser. This increasing density of the bus network also correlated with the citizens' satisfaction with bus transport, which increased with the proximity to the CBD (Wachs 1976: 102).

From the varying density of the bus network alone, it followed that lower-income groups who could not afford a car *had* to also live near the city center, where a local transport system existed. The significant correlation in Los Angeles that low-income households lived closer to the city center and used public transport more often was also confirmed by the national average:

In 1970, 82 percent of all SMSA households with incomes under $5,000 lived within six blocks of public transportation, by comparison with 58 percent of those with incomes over $15,000. (...) In practice, individuals from low-income households relied on transit for 14 percent of their trips, whereas those with incomes over $7,500 used it for only 4 percent (Altshuler 1979: 270).

Only the Los Angeles citizens who owned a car were able to benefit from the mobility gain made possible by the extensive freeway network, for example, to move to places with better living quality. The residential areas of low-income groups showed an impressive continuity throughout the 20th century. The "cheap" residential areas extended like a ring around the center and spread south to the coast. They had the lowest property and rent values and were also described by the population as the worst residential areas (Nelson/Clark 1976: 27 ff).

> Within the confines of this area (...) are the lowest income, most poorly educated families occupying the worst housing. Although the area may be regarded as a totality of poverty, four important ghettos are located here: the Mexican-American settlement centered on East Los Angeles at the northeast; the Negro communities of Florence and Watts at the south; Chinatown at the north; and the no-man's land of skid row as a buffer (Burns/Harman 1968: 67).

The lowest-income groups were largely ethnic minorities, who were spatially isolated, especially among themselves.[13] As far as the settlement of better residential areas was concerned, Anglo-Americans of medium and higher income in particular had benefited from the increased accessibility of the metropolis. For them, automobility brought the opportunity to live in the better residential areas further out of the center, for example, near the beach, in the mountains of Santa Monica in the west, Palos Verdes, and at the foot of the San Gabriel Mountains.

In an exemplary study of mobility behavior, Perloff and others (1973) compared a rich quarter (Beverly Crest) and two quarters with low average incomes (Watts, Boyle Heights) in Los Angeles. It turns out — as was to be expected — that the residents of Beverly Crest on average not only traveled further but also had a wider range of destinations within the metropolis than the residents of Watts and Boyle Heights (see Table 3.5). The car density of households was as follows: While 94% of households

---

[13]The Hispanic neighborhoods expanded with a sharp rise in population to 39%. It was expected that Hispanics would soon outnumber Anglo-Americans, who then made up 50% of the population, and become the largest population group. The expansion of the Hispanic population was increasingly alarming for the socioeconomically worst-off black minority, which now represented only 14% of the population and would soon be overtaken by the growing Asian group (11%). The poor economic situation of the black minority, which was the smallest minority soon to be completely forgotten, and the threat of being displaced spatially by other somewhat better-off minorities were some of the reasons for the racial riots in April 1992, which claimed 51 lives.

Table 3.5: Accessibility of three different neighborhoods in Los Angeles.

Accessibility index by income in SBS	Beverly Crest	Watts	Boyle Heights
Total	5.97	1.71	2.11
0–4,999	2.74	2.22	1.86
5,000–6,999	n.a.	2.12	3.06
7,000–7,999	1.65	1.15	1.28
8,000–8,999	n.a.	0.67	0.77
9,000–9,999	1.72	0.74	1.18
10,000–12,499	4.19	0.88	2.14
12,500–14,999	n.a.	n.a.	0.87
15,000 and above	7.97	0.33	0.35

Source: Perloff *et al.* (1973: 194).

in Beverly Crest owned at least one car and 62% had two or more, only 58% of households in Watts and 60% of those in Boyle Heights had a car available. Their figures for two or more cars were 2.6% and 11.7%, respectively. It followed that 40% of families were totally dependent on the — unsatisfactory — bus system.

Another surprising result of the study was, however, that it was not only the average distance traveled but also the range of destinations within the metropolitan area that was important. Although the accessibility index to workplaces[14] showed, as expected, a significantly higher value in the rich district of Beverly Crest (5.97) than in Watts (1.71) and Boyle Heights (2.11), when broken down by income, the index revealed that the *lower*-income brackets ($0–$7,000) in Watts and Boyle Heights had relative ease in getting to work. This indicates the relative proximity of the lower-income workers to their (lower paid) jobs and it followed that the bus users, who were usually economically worse off, had to live closer to their workplace (Commuter Transportation Services 1992: 18). By contrast, the residents of Watts and Boyle Heights had considerably less access to better paid jobs. The index for jobs above $7,000 slipped sharply

---

[14]The *accessibility index* expresses the relative number of jobs that can be reached within a certain travel time. Not only does the availability of an appropriate infrastructure (freeway connection, bus network) play a role but so does the availability of the means to participate (car, bus fare).

(due to the moderate distance between Boyle Heights and the better-paid (white collar) jobs in the CBD, the index of income of $9,000–$12,500 was somewhat higher there again).

On the other hand, the residents of Beverly Crest achieved their highest accessibility index with high income. This pointed to the relatively fixed structure of the respective neighborhoods in terms of accessibility to jobs and at the same time posed a dilemma for the residents of Watts and Boyle Heights. They had to live in the poorer neighborhoods in order to reach their workplaces reasonably well. And even if they could get a better-paid job, it would be difficult for them to reach it.

The study supported the general observation that whoever could do so moved away from the bad cheap housing areas and preferred to accept the long commute. This explained Perloff's paradoxical result of a higher accessibility index to jobs for lower-than-middle wage groups. However, high incomes usually also had a very high accessibility index. This pointed to a high rate of car usage by households and the proximity of motorways, which allowed a large number of jobs to be reached (cf. Perloff *et al.* 1973: 185 ff).

3.4 Consequential problems of motorized private transport

The construction of the general road network in the 1920s could not keep pace with the increasing number of journeys made by citizens, so traffic problems became more acute. With the disproportionately higher capacity of a freeway network, transport policymakers believed they had found the technical solution to meet the increased demand for mobility after the Second World War. The following section examines the impact of the freeways on the mobility of citizens.

In the context of the presentation of mobility behavior (cf. Section 3.3.2), it was already mentioned that the spatial mobility radius had been extended by the freeways. This was reflected even more clearly in the driving performance of the Angelenos. While the population of the metropolis doubled between 1950 and 1970, the number of miles traveled almost tripled in the same period from 54 million to 145 million miles. Ten years later, the mileage further increased by 50% to over 220 million miles per day (1980), although the population of the metropolis increased by only 11%.

3.4.1 Freeways and mobility development

How did the high traffic capacity of the freeways affect the travel times of citizens, which, according to their supporters (cf. McElhiney 1960), should be shortened? The development of travel time also indicates whether the freeways were able to cope with traffic flows. Studies were initially able to establish that travel time in the metropolis had actually shortened in the 1950s. Despite a threefold increase in the number of cars between 1936 and 1957, travel times between the CBD and 14 places in the surrounding area had been reduced from 33 minutes to 26 minutes on average during normal traffic hours (Koltnow 1970: 6–10, quoted from: Altshuler 1979: 320). However, studies already showed in the 1960s that, during peak traffic, travel times increased again considerably. Within less than a decade, average travel times on five freeways during rush hour rose two and three times. At the end of the 1960s, 25% of the freeways were congested during peak hours (Table 3.6).

The Angelenos were now stuck in traffic jams for about 10% of their 629,000 hours of driving, with over 40 million daily trips (1984). If no action were taken, the Southern California Association of Governments (SCAG) scenario would result in a threefold increase in travel time in

Table 3.6: Development of peak travel times of selected freeways in Los Angeles.

Section	Average journey time (minutes)	
	1956–1958	1965
Harbor Freeway, Civic Center to Santa Barbara Avenue	5:50	10:00
Hollywood Freeway, Civic Center to Hollywood and Vine	7:35	25:00
Pasadena Freeway, Civic Center to Glenarm	10:30	25:00
San Bernardino Freeway, Civic Center to Rosemead	12:51	30:00
Santa Ana Freeway, Civic Center to Norwalk Boulevard	16:40	40:00

Source: McElhiney (1960: 305) and Automobile Club of Southern California (1966).

2010, with citizens spending more than 50% of this time in traffic jams (SCAG 1989b: I-2).

The declared goal of the freeway planners to cope with future traffic flows could not be achieved. The construction of freeways produced unexpected side effects by inducing additional traffic in the *short* and *long terms*. The traffic induced in the short term came from two sources: on the one hand, journeys previously made using other modes (e.g., bicycle, bus) and, on the other hand, journeys that were not made at all. In fact, a traffic study in Los Angeles calculated that 29% of the traffic on the Santa Monica Freeway was induced traffic that was created shortly after its opening (McDaniels 1971: 21 f). Although traffic was reduced after the opening of new freeways built in parallel streets, traffic censuses showed that traffic was heavier there years later than before the freeway was built. Access roads already had to cope with massively increased traffic flows when freeways were opened. They later led to the installation of traffic lights at freeway entrances (*ibid.*: 11 f).

In the longer term, the volume of traffic will increase as a freeway affects the mobility structure and generates new settlements. The indicator of this process is the increase in land prices along the freeway corridor with the publication of the planned routes. As the proximity to the freeways improves transport connections, areas further away from the workplace can also be settled.[15]

The effects of automobility in Los Angeles are made particularly clear by comparing other cities. The different mobility behavior patterns are already evident in a comparison with Boston, a city with a better public transport system, fewer freeways, and less cars per capita (average weekdays, 1970):

- A resident of L.A. makes 32% more trips (motorized transport only) per day, using the car 44% more often than a resident of Boston.
- At 19.1 miles, the average trip length in Los Angeles is almost twice as long as in Boston (9.9 miles), but the average speed of the car is also almost twice as high (36.5 mph vs. 19.1 mph). As a result, the

---

[15] It is irrelevant to individual decisions that, especially during peak periods, travel times will later slow down again. In fact, the opening of a new section of road will first relieve the traffic situation. This is the decisive factor in the decision. Other factors also play a role. For example, the desire to live in low-density areas inevitably drives the decision to move to the surrounding area. A possible congestion of the route is usually accepted.

average travel time by car for citizens in both regions is about the same.

• The Angeleno spends on average 46% more time in the car than the Bostonian (42.2 minutes vs. 28.9 minutes). While the former makes only 1.6% of the trips by public transport, the latter does so with 10.4% of the trips.

• Overall, a citizen of L.A. spends on average 20% more time in traffic than a citizen of Boston (43.3 minutes vs. 36.1 minutes daily), even though the Angeleno has a far superior freeway network (cf. Fauth *et al.* 1975: Table 1, quoted after: Altshuler 1979: 324 f).

This comparison makes it clear that the higher speeds made possible by cars and freeways did not reduce the daily travel budget (time saving), but rather encouraged longer journeys, which can mean considerable *time loss* at peak hours. Conversely, the lower travel speeds of Boston's citizens (see L.A.) had not led to longer travel times.

3.4.2 Freeways and traffic safety

A further argument in favor of urban motorways is that they are safer (McElhiney 1960). If one takes the size of *mileage* preferred by traffic engineers as a reference, then the freeways really do have a positive effect. Whereas in 1939 there were 9.5 fatalities per 100 million vehicle miles, the rate in 1957 was only 3.4. Compared to the normal road network (3.4), the freeways also had a better accident fatality rate (2.4) in 1957 (Nelson 1983: 284).

However, the positive balance of the freeways in terms of road safety is only valid when measured by the degree of risk per mile driven. McDaniels pointed out that individuals are less interested in the risk per vehicle mile than in the risk of their total journeys (1971: 27 ff). If the *personal risk of a citizen in traffic is* expressed in terms of "annual accident fatalities per 100,000 inhabitants", a different picture emerges (cf. Table 3.7).

In a metropolitan comparison, Los Angeles had by far the highest accident risk. Since the freeway was built in the early 1950s, the figures have not shown any significant improvement. The safety advantage of the freeways per mile driven had been compensated by much higher mileage. It can be assumed that the more than 50% higher accident risk in

Table 3.7:   Road deaths per 100,000 inhabitants.

Year	New York	Philadelphia	Chicago	Los Angeles
1942	10.8	12.7	13.7	21.3
1945	9.6	9.1	13.9	27.9
1947	9.6	10.2	15.0	21.8
1948	7.5	7.3	12.5	15.5
1951	7.0	8.0	11.5	13.6
1954	7.5	8.9	10.9	14.8
1957	8.1	8.2	8.5	17.8
1960	7.6	8.5	7.6	15.7
1963	8.5	8.7	8.2	13.6
1966	7.4	9.3	9.0	16.2
1969	10.7	10.0	9.4	17.2
1972	10.2	11.7	8.4	15.4
1974	9.0	9.3	8.9	11.3
1975	n.a.	8.9	9.7	9.7
1977	8.6	6.5	9.2	12.2
1978	8.4	6.5	9.1	14.2
1980	9.0	7.5	10.0	15.7

Source: National Safety Council, Accident Facts, annual expenditure.

Los Angeles compared to cities of similar size was due to the higher density of cars. For example, in 1970, the metropolis of L.A. had the highest number of cars per household (1.33 cars/household) compared with New York (0.75), Boston (1.07), and Philadelphia (1.13) (Nelson/Clark 1976: 56).

As can be seen from the table, the city's traffic risk did not decrease until 1974, only to increase again thereafter. This could be related to the fact that a speed limit of 55 mph was introduced on freeways in 1974 (cf. also the sharp fall in the accident rate on freeways per 100 million miles traveled in 1974, in: Howell/Cutler 1990: 8). Another possible explanation could be the fuel price increases and reductions in mileage associated with the oil crisis. This would also make the rise in the accident rate in subsequent years more plausible. Finally, it should be noted that in 1990 the California Department of Transportation still used *car-friendly* mileage as a reference value for the accident rate (Howell/Cutler 1990).

In summary, the freeways have not been able to solve the mobility problems of the Angelenos despite the huge investments made since the 1950s. The provision of transport capacity was continuously (over)compensated by increasing demand, so the quantity but not the quality of the infrastructure services increased. In addition to the high risk of accidents in the metropolis, poor air quality in particular is perceived as a negative side effect of car ownership. In the introduction, it was mentioned that in 1990 the metropolis of Los Angeles violated the federal ozone standards three times as often as the region with the second worst ozone standards in the US (cf. SCAQMD 1991a, 1991b).

Chapter 4

The Return of Rail Transport
to the Metropolis

Leading the way to greater mobility

Slogan of the Los Angeles
Transportation Commission

This chapter outlines the most recent phase of transport policy development in Los Angeles to date, which led to the reintroduction of rail-based
public transport in the metropolis. Subsequently, the more recent transport
policy concepts that are intended to solve the mobility problems of the
metropolis are outlined. First, however, the development that led to the
construction of a rapid transit system is briefly described.

4.1 High-speed railways as a solution to mobility problems

Poor air quality in the region and increasing traffic problems have contributed to a steady decline in public support for freeway construction since
the late 1960s. Even in suburban areas of the metropolis, such as Orange
County, traffic had for years been regarded by the population as by far the
most important public problem (Baldassare 1991: 212). Since the 1970s,
the construction of freeways slowed down considerably, as the funds
available for financing shrank in real terms and, in addition, many projects
failed due to the resistance of municipalities and civic organizations.

Since that time, there has been renewed talk of a "traffic crisis" in the metropolis, as the rate of traffic growth has considerably exceeded the already high population growth rate (Wachs/Guiliano 1992b: 3).

After the discontinuation of rail-based local transport at the beginning of the 1960s, the bus was the only public transport mode, but it could hardly contribute to an improvement in the transport situation. The operation of bus lines along the former route of the tram and regional trains after the Second World War by private transport companies was very sparse and, moreover, did not follow the settlement patterns of the metropolitan population. This led to considerable public dissatisfaction and a further relative loss of passengers. After the Southern California Rapid Transit District (SCRTD) had taken over most of the public transport system, some of the buses, which at the time were among the oldest in the US, were replaced by new ones with state support over a relatively short period of time (cf. OECD 1988: 84 f). Nevertheless, the importance of local public transport remained negligible. In 1980, for example, of 39 million passenger trips per day, only around 1 million were made by bus (*ibid.*: 83).

In view of urgent traffic and environmental problems, there have been increasing attempts since the late 1960s to bring back *rail-bound* public transport to Los Angeles. In 1964, the newly created SCRTD replaced the Los Angeles Metropolitan Transit Authority (LAMTA), which in 1958 operated most bus routes and the remaining rail networks as the public transport company of Los Angeles County. Like the LAMTA, the SCRTD was responsible for operating the bus lines and planning and building a rapid transit network. Unlike its predecessor, however, the SCRTD had greater powers and was able to collect taxes and issue bonds as an independent administrative unit — with the consent of the population.

In 1968, an elaborated plan for a rapid transit system was submitted to the population for approval. It consisted of an 89-mile two-lane network with five corridors and some 20 miles of underground tracks. In addition, 115 new bus lines were to be established to take passengers to the stations. The total cost of the entire system was estimated at $2.5 billion, to be financed by a 1 1/2% increase in VAT. In the referendum, however, only 45% of voters approved the proposal (Nelson 1983: 292 ff).

In 1974, a second attempt to build a rapid transit system also failed. The plan envisaged a 145-mile rail network, supported by 1,000 new feeder buses, at a cost of $4.5 billion that was to be financed by a 1%

increase in VAT. This proposal was also rejected by L.A. County voters, with 43.3% in favor and 56.7% against (*ibid.*: 296 f). Two years later, the electorate finally rejected a third proposal, which envisaged the construction of a much larger rapid transit system, but completely neglected the bus lines (*ibid.*: 297 f).

It was not until November 1980 that L.A. County voters approved the Los Angeles County Transportation Commission's (LACTC) proposal. The plan of this new authority, which was set up following the third rejection of a rapid transit network (Weekly 1990) and which had even greater powers, provided for an increase in VAT of half a percentage point (Proposition A), which was intended, among other things, to finance a rail-based public transport system for the metropolis. However, in order to secure the approval of the citizens for the increase, the LACTC made great efforts and very far-reaching concessions to various groups within the county (cf. Section 6.2.). Finally, in the election on 4 November 1980, 54% voted for Proposition A (Nelson 1983: 150 f). The approximately $400 million available annually through these tax revenues was to be used from then on — in addition to improving bus operations — to finance, among other things, the construction of a 160-mile-long rapid transit system. In 1990, voters agreed to a further increase in VAT by half a percentage point for local public transport (Proposition C), making a further $400 million available annually starting in April 1991 (Weekly 1990: 5 f). Together with other federal and state funds, the huge sum of over $183 billion was to be spent over a *30-year transport plan* to improve the mobility of the population (LACTC 1992).

4.2 The 30-year transportation plan

The transport policy in Los Angeles has been focused on the car since the 1920s. The answer to mobility problems was mainly to increase the capacity of the road or highway system. In contrast, the expansion of the capacity of the public transport network, especially of the *rail-bound* local transport, was to be given priority over other measures. Referring to the 30-Year Transportation Plan, Wachs and Guiliano stressed that "(T)he recent change in direction in transportation policy is one of historic proportions, which attempts to reverse a trend which has been powerful for over sixty years" (1992b: 3). The implementation of this new approach would exceed, adjusted for inflation, the annual expenditure spent on building freeways during peak times (*ibid.*: 1992b: 5).

The car as a mode of transport played only a marginal role in the new program. It would only fill some gaps in the freeway network and make technical improvements to *existing* freeways. The "highway component" of the 30-year plan included increasing the capacity of the freeways by installing over 300 miles of carpool lanes and technical traffic management systems (e.g., ramp metering). According to the "bus component" of the plan, the city intended, inter alia, to purchase over 100 buses a year during the first six years and to operate an additional 1400 buses at peak times (LACTC 1992). The measures operating under *Transportation Demand Management* represent an interesting new component of transport policy in Los Angeles, which is outlined in more detail in Section 4.2.2. One of the aims of these measures was to reduce the number of commuter journeys by 2010 by 2.4 million per day from 40 million at the time of implementation. This was mainly based on the establishment of car and van pools but also on telecommuting (cf. Section 5.2). However, at the heart of the plan was the construction of a 400-mile rail network consisting of 200 miles of high-speed railways (metros, light rail) and 200 miles of regional railways (commuter rail).

4.2.1 The construction of an express and regional railway system to solve the transport problems: A preliminary assessment

The "railway component" ate up the lion's share of expenditure on the metropolis's transport plan. Over the next 30 years, a total of over $78 billion was to be spent on the construction and operation of a 400-mile (over 650 km) rail network for the metropolis (LACTC 1992: 60). Several sections were opened by the end of 1992. The *Blue Line*, which was opened in July 1990 as the first light rail system, and the Commuter Rail, which started operating in October 1992, were used to make a first preliminary assessment of their potential impact. The main focus was to be on the extent to which these projects influenced mobility problems.

Since July 1990, the 21-mile *Blue Line* has been connecting downtown L.A. with Long Beach in the south of the city. In February 1991, the line served 22 stations with the completion of a tunnel in the city center. In the meantime, the line carried a good 30,000 passengers daily. Although this passenger volume was hailed by both the LACTC and the press as a success that "far exceeded expectations" (Los Angeles Times, 21.1.1991), sober analyses of *real* passenger growth relativized this assessment considerably. Richmond, for example, pointed out that before the *Blue Line* began its operation, a local bus line connecting downtown Los Angeles with Long Beach carried over 31,000 passengers a day (Richmond 1991:43). The passenger volume of the new light rail line could hardly be considered a great success for this reason alone.[1]

Even the sections of the *Metrolink* commuter railway system that opened at the end of October 1992 and connected central L.A. with the communities in the region (cf. Southern California Commuter Rail 1991) only marginally relieved the traffic situation. After the first fare-free week, the new commuter lines, with a current rail network of 112 miles, carried only about 3000 passengers daily after the first weeks. This was only slightly more than a single freeway lane could handle within 90 minutes (Wachs/Guiliano 1992b: 9).

One reason for the low transport figures was certainly the mobility structure of the metropolis, which was characterized by a comparatively low population density (cf. above). In addition, in some cases, including *Blue Line* and *Metrolink*, the route was not primarily selected according to the criterion of the greatest need for transport, but often because the rights of way were acquired as cheaply and easily as possible. For example, the largest part of the *Blue Line* was built on tracks that were used by Pacific Electric's regional railways until 1961 and were then subsequently used for freight transport. However, the settlement and workplace structure in the region changed considerably during the car phase, so efficient utilization of the various lines was hardly to be expected. Negotiations concerning the acquisition of rights of way were nevertheless difficult in the case of the *Blue Line* and made construction more expensive (McSpedion 1984: 426 f).

In fact, the new railway systems did not perform well under *efficiency criteria*. The capital costs of the *Blue Line* alone ultimately totaled almost one billion US dollars (Richmond 1991: 65). Even if these costs were not included, Richmond calculated that fare revenues covered at best 16% of operating and maintenance costs (*ibid.*: 70; Wachs/Guiliano (1992b: 8)

---

[1] In an interview, urban planner Martin Wachs expressed a similar assessment. According to his calculations, only half of the daily passengers previously traveled by car (interview, on 20.11.1992 in Los Angeles).

came to only 11%). The low efficiency of the railways was illustrated by a comparison with the capital and operating costs of bus lines, which in Los Angeles covered on average about 40% of their costs through fares. According to Richmond's calculations, each passenger on the *Blue Line* would at best have to be subsidized with $3.22, while the public authorities would only have to contribute $0.78 per trip for a bus passenger (for a corresponding route) (1991: 72).

The discrepancy in costs between train and bus was likely to be even more significant for commuter railways. Although it was too early for a final assessment, some trends were already visible. The approximately 3,000 daily passengers on the three lines that had opened up at this point did not meet expectations by far. In 1991, the forecast for the start-up phase for just two lines alone was over 6,000 daily passengers[2] (Southern California Commuter Rail 1991: 42). In the first year, the passengers forecast was to cover just 10% of operating costs (*ibid.*: 59). In view of a considerably lower passenger volume, an even lower rate of cost coverage could be expected.

Like the railway system of the 1920s, a large part of the lines of the 30-year plan were designed radially toward the center, although — as described — the jobs in the metropolis were now spread across the entire area (Figure 4.1). The *Blue Line*, the *Red Line*, which was partially redeveloped at the beginning of 1993, and the Metrolink commuter lines all ran toward the CBD, where — with a decreasing tendency — less than 5% of the total number of jobs in the region were located. In addition, the *commuter rail lines*, as their name suggests, were primarily designed for daily commuting from home to work. Accordingly, the trains only went into the city in the morning and back to the suburbs in the evening, so shopping trips with them were hardly an option.

Regional railways were also questionable under *equity criteria.* Mostly middle- and upper-class citizens lived in the suburbs of the metropolis affected by the lines, and the relatively small number of them enjoyed the benefits of an expensive commuter network. The lower classes, on the other hand, benefited far less since they lived near the center anyway. However, it was precisely these people who were far more dependent on public transport. An alternative use of expenditure on improving inner-city bus services would benefit a larger and more needy

---

[2]The third line (San Bernardino-Los Angeles) was only partially opened. Passenger numbers were estimated at 4,490 daily for the completed line.

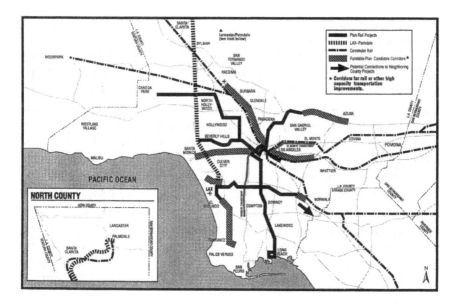

Figure 4.1: Railway projects and planned infrastructure corridors in the metropolis.

Source: Los Angeles County Transportation Commission (1992: 48).

section of the population, using far fewer financial resources (cf. also Wachs/Guiliano 1992b: 10 f).

4.2.2 New transport policy concepts

In addition to the reintroduction of a rail-bound public transport system, the measures circulating under the term *Transportation Demand Management* represented a new dimension of transport policy in Los Angeles. In contrast to the other measures, they did not focus on the transport mode or supply side. The core of the measures was aimed at increasing the proportion of car-pooling for daily commuting to work. The 30-year plan provided for a total of 3% of funds for this purpose (LACTC 1992: 60).

The most interesting of these was the so-called *Regulation XV*, which was adopted due to continuously deteriorating air quality in 1987 by the South Coast Air Quality Management District (SCAQMD) and whose implementation began in July 1988. This regulation required all employers with more than 100 employees (companies, government agencies, schools, hospitals, etc.) to submit a plan within a certain period to increase the average vehicle ridership of their employees (hereinafter AVR: average vehicle ridership) to reach the specific value approved by the SCAQMD. The AVR was calculated by dividing the number of employees by the number of motorized transport vehicles they used. The target value of the AVR was based on population density. In densely populated areas, it was 1.75 workers per car, while sparsely populated regions had to achieve an average of 1.3. In the mid-1990s, the AVR for the region as a whole was expected to be 1.5, although before the start of the measures, this value was still between 1.1 and 1.2 (Guiliano/Hwang/Wachs 1992: 3).

Under Regulation VX, each company had to employ a trained employee transportation coordinator who was responsible for the programs designed to encourage employees to use modes other than driving alone to work (Wachs/Guiliano 1992a). In June 1992, there were some 6,200 companies, public authorities, and other institutions in the region covered by the scheme, which together had an estimated 2.26 million employees. This was about 40% of the SCAQMD's workforce (Guiliano/Hwang/Wachs 1992: 2 f). The implementation of the plan worked with negative and positive incentives such as preferential parking for carpooling, subsidies, or even parking fees. Violations of Regulation XV were to be severely punished. The most common reason for a violation was the simple failure to draw up an implementation plan. The highest fine of $150,000 had been imposed on a retail company (*ibid.*: 4).

Initial investigations showed a measurable impact of the scheme on the mobility behavior of the companies concerned. According to a survey, the AVR increased statistically significantly in the region from 1.22 to 1.25 within one year. Among the companies in the sample, single drivers fell from 75.7% to 70.9% (Wachs/Guiliano 1992b: 12 f). This decline reflected a higher proportion of car-pooling, as other modes, such as public transport, cycling, or walking, did not show a significant increase.

What were the effects of the positive outcome of this measure on the general traffic situation? According to a calculation by Wachs and Guiliano, the traffic volume was to be reduced by 1.3 million vehicle miles per day, which represented an annual reduction in traffic volume of 0.4% (1992b: 13). The impact was so small because commuting to work accounted for only about a quarter of all journeys and only just under 40% of employees worked for companies with more than 100 employees (Christiansen/Gordon 1992: 6).

Although largely employee supported, Regulation XV was not particularly cheap. A survey for the SCAQMD of 5,763 regulated companies and institutions found annual costs of $105 per employee arising from the measure. Eliminating a single-vehicle trip cost $3,000 per year (Eamst/Young 1992).

In addition to Regulation XV, the *Traffic Reduction and Improvement Program* (TRIP) adopted by L.A. City Council was also an attempt to reduce the demand for transport. Under this framework law, the City Council could declare a district or subdistrict a "traffic impact area" by a 2/3 majority. This made the area subject to special land-use controls and development charges, which were intended to alleviate the (negative) traffic impact of new development projects. For example, *developers* of new office space (above a certain size) would have to determine the additional car trip rate caused by their project and take measures (e.g., car-pooling, job tickets) which would reduce the induced traffic for the project concerned by 15% during peak afternoon traffic hours. If this was not achieved, a one-time fee (between $2,000 and $6,000) for each unreduced journey had to be paid into a fund to finance traffic reduction measures for the area concerned (Wachs 1990. 264 f).

In addition, the so-called *Proposition U*, approved by the voters of the City of Los Angeles in 1986, sought to directly influence mobility structures to prevent traffic. By limiting the amount of space available for offices through zoning, a more balanced mix of uses was to be achieved. Although the measure explicitly exempted various business centers from "downzoning", the ratio of commercial to residential space for new buildings was halved from the previous 3:1 to 1.5:1 (Wachs 1990: 245).

These new transport policy action patterns will be briefly discussed again in the next chapter in a comparison, in which their effects on mobility patterns will be assessed (Section 5.2).

Chapter 5

Mobility Patterns and Public Policy in Comparison

Analytic activities have tended overwhelmingly to focus on the appraisal, advocacy, and/or incremental adaptation of (...) technologies and services — which we term preselected solutions — rather than on laying bare the character of the problem generating demands for public action or searching with a fresh eye for effective remedial strategies. Paramount among the preselected solutions have been highway and transit improvements, and policy discussion has typically proceeded as if these were the only options available for addressing sources of dissatisfaction with the urban transportation system.

Alan Altshuler (1979: ix)

For the different phases of Los Angeles' transport policy development, the previous chapters have analyzed mobility patterns and public policy action. The aim now is to compare the mobility patterns and transport policies of these periods in a longitudinal section and to systematically present their similarities and differences. To this end, reference will first be made to the thesis formulated at the beginning of the paper and the question will be asked about what mobility patterns are formed by the two motorized modes of transport — rails/transit and freeways/car. Then, the focus is on the interrelation between mobility patterns and transport policy action. In the first step, the dominant (material) political actions and strategies described above are typified according to their transport policy orientation. In the second step, an attempt will be made to evaluate the effects that transport policies have had on mobility patterns.

5.1 The mobility patterns of the railway and car metropolis

First, it is important to point out possible methodological imprecisions, which may result mainly from the lack of exact data (on mobility behavior, for example) in the first decades of 20th century. Nevertheless, the empirical data used here are sufficient to identify the similarities or differences in mobility patterns that have emerged during the various phases of the city's transport policy development.

Trams and regional trains were the dominant mode of transport in Los Angeles during the first phase from the end of the 20th century until after the First World War. In the 1920s, the automobile had already started to break the dominance of the railways as a means of transport. In the following period (2nd phase), the automobile had risen to the dominant motorized mode of transport in the metropolis. The automobile retained this monopoly-like position until today. The comparison of these two phases is now carried out based on the criteria of mobility structure, mobility behavior, and mobility opportunities.

5.1.1 Mobility structure

The dominance of rail transport in the first phase and of the automobile in the second phase had similar effects on the geographical location of the sites of activity. Surprisingly, both the railways and the automobile supported the tendency toward urban sprawl, albeit to varying degrees. Pacific Electric's regional railways were pioneers in creating the conditions that enabled citizens to live in the suburbs of Los Angeles. Until the First World War, the electric railways had a de facto monopoly on local transport. The profitable intertwining of passenger transport and real estate speculation led to an increasingly *horizontal* settlement structure, as more and more citizens of the metropolis took up the offer and settled far outside the city center. As early as 1920 — when the railway was still the dominant means of transport — almost half of the population lived outside the central areas of the city. Admittedly, the Angelenos' dependence on the railways only led to a limited growth in urban sprawl. Although widely dispersed throughout the region, the layout of the lines was relatively rigid, with houses within walking distance of the railway stations.

Freed from such obstacles by the new and much more flexible automobile transport mode, the horizontal settlement structure created by the railways increased exorbitantly. The dispersal of newly built dwellings in the metropolitan area — with growth rates of over 600% in the 1920s — reached undreamt of levels in the following decades. Considerable population growth and the construction of freeways since the 1950s made Los Angeles an extreme type of settlement area characterized by low-density and single-family houses (Wachs 1984: 297). The differences in the impact of the rail and car modes of transport on the settlement structure are therefore only of a quantitative nature: the sprawling structure of the metropolis had already been created by the regional railways and was only considerably strengthened by the new mode of transport, the car. Contrary to what is often said, rail-bound public transport does not *per se* lead to dense settlement structures.

Until the end of the 1920s, most jobs and consumer locations were still located in or near the CBD. This meant a considerable concentration of destination traffic, especially during peak traffic periods. Since until after the First World War the electric railways were virtually the only motorized means of transport, their rail network, which was laid out radially toward the center, structured the traffic flows. Trade and industry initially saw no reason to settle outside the center. On the one hand, the potential demand was usually very low due to the low population density in the surrounding area and, on the other hand, the radial network of the railways ensured good access for the population to the CBD until the end of the First World War, while accessibility for workers and customers was far worse due to the lack of flexible means of transport in the surrounding areas. Finally, the companies also lacked satisfactory means of freight transport. Until after the First World War, goods were transported to the city by horse-drawn wagons, which in turn prompted entrepreneurs to locate near the large freight stations in the center or at the port.

The 1920s marked the turning point in this development, as all three factors changed during this period. With the population boom of the 1920s, the population density also increased in the surrounding area, making these areas more attractive for retail trade. For example, while the San Fernando Valley and Glendale had a population of around 32,000 in 1920, the population had risen to 205,000 ten years later. Twenty years later, the population there had tripled again (cf. Pegrum 1968: 564). While the inaccessibility of the center came to a crisis-like climax in the 1920s, the spread of the car improved the attractiveness of industrial and consumer locations in the region, as they became increasingly accessible for workers and customers. Finally, the truck (HGV) established itself in the economically important period after the First World War, when many industrial companies settled in the region. The lorry gave especially space-intensive branches of production the opportunity to settle on cheaper land outside the center, while still enjoying the advantages of more efficient ground-level production.

These developments in the 1920s led to an increase in trade and industry in the surrounding area. Only the financial sector and other services were still largely located in the center. Although the CBD still provided the largest number of jobs, the tendency toward dispersion was more than evident. Only about 1/3 of the jobs in the region were still located in so-called *subcenters*, while 2/3 were scattered across the metropolis.

In terms of the geographical location of workplaces and places of consumption, the rail and car modes of transport did not therefore tend to develop differently either. Rather, the conditions for the establishment of retail and industrial companies seemed to depend primarily on factors such as population density (and consequently purchasing power potential) and spatial accessibility. Which mode of transport ensures accessibility is a secondary question. As with the settlement structure, the flexibility of the car led to a higher degree of dispersion than would have been possible with the railways. The reach of rail travelers was limited to a fairly limited radius around the stations.

On the other hand, the dispersion of jobs in the auto phase did not at all lead to mixed-use structures. On the contrary, the sharp separation of residential and working uses, which was already clearly established in the railway phase of the metropolis, continued in subsequent decades. While at the beginning of 20th century the regional trains of *Pacific Electric* allowed long commutes from home to work, the successive construction of a freeway network, especially after the Second World War, which in many places followed the former routes of the railways, created the conditions for motorized private transport to be able to cover long distances to work within a bearable travel time. While in the first two decades of the 20th century the lines of the PE and the LARY were overcrowded and progress in the CBD was sluggish, L.A.'s congestion on the freeways is characteristic of the car phase. The reasons for this are the direct accessibility of most destinations via the freeways.

However, while the city's rail phase was characterized by at least a moderate density of source traffic and a high density of destination traffic (in the CBD), a thinning of source and destination traffic can now be observed. This is an expression of the abovementioned (increasing) dispersion of jobs and the continued specific "low-density" lifestyle of the population, which preferred living in remote areas to dense housing structures (cf. Section 6.3.2).

The mobility structure has a strong momentum of its own. While at the beginning of 20th century the railways only made it possible to live far away from places of work and consumption, these patterns have become independent. In the meantime, the possibility of mobility has changed unnoticed into a mobility necessity due to the dynamically developing separation of uses and the low-density lifestyle. In this sense, the individual has lost his or her freedom of action. Even if the population were to question the mobility structure, their scope for decision-making would be severely restricted. For example, if a family wanted to live closer to their workplace, they would generally have to give up their preferred and usually relatively cheap housing in the often-remote areas of the metropolis. The mobility structure that is systemic in this way (especially the low population density) also has consequences for transport modes and mobility behavior. It tends to encourage the use of private transport and increases the probability of longer journeys.

5.1.1.1 *Mobility behavior*

If one compares the railway and car phases regarding the mobility behavior of the Angelenos, many similarities become apparent. For example, the monopoly-like dominance of the respective transport mode is striking. The citizen has no transport alternatives, especially since the distances between the places of activity are for the most part too long for walking or even cycling. Since the metropolis has a comparatively low population density (in 1920 — before the population boom — it was only 646 inhabitants per sq. mile), most of the distances cannot be covered on foot.

In particular, the high transport numbers indicate the dependence of the Angelenos on the railway companies. A 1911 traffic study calculated that the inhabitants of L.A. used the tram twice as often as citizens of other cities of comparable size (Bottles 1987: 33). In the 1920s, the automobile broke the dominance of the railway as the mode of transport in the metropolis. According to a transport study, the automobile had already taken over the dominant position in local transport in most areas of the metropolis in 1941 — while the railway was still in operation in L.A. — and in the years that followed, it considerably expanded this position with the construction of the freeway. Nowadays, an average of around 85% of all commuting to work is done by car, although this proportion is much higher in more remote areas.

The separation of residential and workplace use was also designed in the railway phase. The distances between the places of activity of citizens (distances between routes), which were already greater than in other cities at the time, increased again in the following decades. In comparison to other metropolises, all income groups in L.A. today travel considerably further than their counterparts in other US metropolises. The extensive rail network of the 1920s and the freeways of the 1950s and 1960s increased the (potential) accessibility of the metropolis. In terms of average travel/commuting times, no figures can be given for the rail phase, although the sharp separation of use is likely to have had a negative impact on travel times. However, a comparison of transport budgets between Boston and Los Angeles shows that at 43.3 minutes per day, the Angeleno spends 20% more time in traffic than the Bostonian. The most recent travel time studies for commuting to work also show that Angelenos, at 36 minutes on average (1992), commute much longer than the national average of 22 minutes (cf. Section 3.3.2).

5.1.1.2 *Opportunities for mobility*

Since at the beginning of 20th century no data on mobility behavior were collected separately for different socioeconomic groups, a direct comparison of mobility opportunities is difficult. However, if one asks which social groups benefited from the mobility offered by a better residential area, statements can be made. The material analyzed here shows that railways and cars have similar effects on the mobility opportunities of social groups. Both the railways and the car enabled middle- and upper-income citizens to distance themselves spatially from less-well-off socioeconomic groups. The latter could only benefit to a very limited extent from the general increase in mobility. However, the transport modes were only the *means* of achieving spatial distance from undesirable factors such as poverty, crime, ethnic heterogeneity, dirt, and narrowness. The reasons for this can be found in corresponding values and attitudes (cf. also below: Section 6.3).

The (spatial) dividing lines in the metropolis run between socioeconomic and ethnic groups. Thus, only the so-called White Anglo-Saxon Protestants (WASPs) of the socioeconomically better-off upper and middle classes benefited from the increased mobility offered by the railway lines. This gave them the opportunity to move to the better residential areas of the suburbs. In contrast, blacks and Latinos were prevented from using the mobility made possible by the railways to live in the better areas outside the center. Planning restrictions as well as simple expulsions by the white population led to the described ghettoization of the different ethnic groups. Latinos and blacks in particular moved into the vacated white flats in or near the center, mostly as tenants.

This pattern was continued with the new mode of transport, the car, and with its spread, "resident segregation" (Park 1925) was also increased. The potential accessibility of better residential areas by car hardly changed the factual (spatial) isolation of ethnic and socioeconomic groups. It continues to this day, although only a good 11% of households have no car at their disposal. Income and skin color still determine the quality of the place of residence.

The potential accessibility of jobs was less asymmetrical in the railway phase than in the car phase of the city. Nevertheless, poorer classes were able to use the relatively cheap trams and regional trains to get to their jobs. However, they lived significantly closer to their jobs anyway, so the upper and middle classes of white Angelenos tended to benefit from the regional railways in this respect. Of course, the potential accessibility of a large number of jobs in the center did not change the fact that blacks, Hispanics, and poorer classes continued to do the lower-paid jobs.

With the dominance of the automobile as a mode of transport after the Second World War and the decline of the railways and the public transport network in general, mobility was increasingly linked to owning a car. As the new mode of transport increased the distance-intensive mobility structure, citizens without a car became increasingly immobile. In Los Angeles, mobility was and is inextricably linked to automobility. Without a car, the already socially disadvantaged were also spatially isolated from jobs and other social groups. Ironically, even freeways, which are supposed to make mobility possible, are the concrete partitions of social classes. The Harbor Freeway, for example, cuts off the socioeconomically lowest strata to the east from the middle classes living to the west (cf. Brodsly 1981).

Despite the already considerable disadvantages due to poorer education, prejudices, and the like, these disadvantaged groups lacked the opportunity to get a better job and thus improve their social situation due to a lack of means of transport. The *McCone report* following the race riots certainly failed to address the lack of mobility of blacks as one of the *causes* of poverty. After all, even when they owned a car, job opportunities for blacks and Hispanics were very limited. It is true, however, that a poor public transport system can have a reinforcing effect in this respect. Moreover, the subjective feeling of immobility and the perception of being trapped can also generate frustration and increase the propensity to violence. This was certainly a factor that contributed to the riots in Los Angeles in April 1992 (*Time Magazine*, 18.5.1992).

It is worth mentioning the observation of Perloff *et al.* (1973) that poorer classes often have easier access to their jobs than, for example, white middle-class people. This observation can be explained by the fact that the former live closer to their — lower-paid — jobs, while the latter, although living in slightly better neighborhoods, have a much longer distance to their jobs. The scarce economic resources of poorer strata of society therefore necessitate a more sensible and ecological lifestyle from a transport perspective (cf. also the remarks on the ecology of scarcity in: Prittwitz/Wolf 1993).

In transport statistics, mobility was largely linked to motorized modes of transport, which meant that the feet and the bicycle as a mode of transport were not included. However, if we measure mobility in Los Angeles in terms of the *number of daily trips*, it is generally accepted that as household incomes rise, so does mobility. Low-income households often make only 1/3 of the trips made on average and travel much shorter distances. As Altshuler (1979: 269) observed, car ownership changes mobility behavior dramatically, so families with one car and more make (car) trips much more frequently and cover considerably longer distances. Low-income households, on the other hand, tend to make far more journeys by public transport.

5.2 Political patterns of action and their effects

Against the background of the transport policy development in Los Angeles, the dominant transport policies are typified in the following in the first step. The effects of these patterns of action are assessed in the second step with a view toward finding a more sustainable solution to the problems of mobility.

5.2.1 Dominant transport policy action patterns

The dominant physical transport policies in L.A.'s development can be typified according to the variables on which the measures are aimed. In turn, the mobility concept serves as a frame of reference.

First, material transport policy action in the broader sense can be aimed, on the one hand, at changing the mobility structure. *Mobility structure-oriented* measures aim to change the spatial structure of living, work, consumption, and leisure. The planning of mixed-use structures and the densification of activity locations are possible strategies for avoiding traffic (cf., e.g., Cervero 1988, 1989a, 1989b, 1991a, 1991b; Wachs 1990).

Second, policies can also address the mobility behavior of individuals. Since they aim to influence individual demand behavior (e.g., changing modes of transport, avoiding peak traffic), they can also be described as *demand oriented* (Guiliano 1988: 156 ff). Subsidizing or taxing certain modes of transport, supporting car-pooling, and putting in place persuasive policies are strategies that attempt to influence individual behavior.

Third, policy measures can also be modal or *supply-side oriented* by directly influencing the supply of motorized transport. Supply-led public action can be distributive, by making areas more accessible to transport across wider regions (e.g., extending motorways over less developed areas). However, supply-oriented measures can also target the capacity or transport performance of transport systems (cf. Figure 5.1). In addition, a negative supply orientation of transport policy is also conceivable by reducing the opportunities for traffic movements. Examples of this would be, for example, traffic calming in residential areas, road dismantling, reduction of the public transport network, or of the frequency of public transport services. Finally, political action can also aim to improve the mobility opportunities of social groups. Reduced public transport fares for

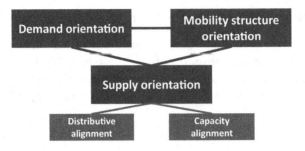

Figure 5.1:   Transport policy types of action.

certain groups and the introduction of a bus line in Los Angeles for the black population in Watts are possible political strategies. However, since the *dominant* actions are primarily examined in our context, we will not go into this in detail.

According to this typology, the development of the transport policy in Los Angeles is predominantly mode or supply oriented. Following the periods when the metropolis had serious traffic problems, the predominant measures were all aimed at increasing the supply of motorized modes of transport. Both the General Road Transport Plan of 1924 and the Freeway Plan and its successive implementation after the Second World War were aimed at increasing the capacity of the (road) transport system and its spatial extension (distribution) throughout the metropolis in order to satisfy the demand for mobility. Even the 30-Year Transportation Plan of 1990 is — measured in terms of expenditure — predominantly supply oriented. However, in contrast to the earlier car-oriented approaches, this plan focuses on solving mobility problems by increasing the capacity of the public transport system.

Characteristically, the strategy of private transport companies (until about 1913) was to respond to rising transport volumes by constantly increasing their capacity (e.g., higher cycle times) and expanding their route network. Intertwined with real estate speculators, they were profit-oriented companies interested in their customers covering long distances (with their railways) every day.

The policies of the following years were remarkably like those of the railways. Similar to the railways, attempts were made to meet the increasing demand for mobility by improving the range of transport modes on offer.[1] In contrast to the railways, a road network was now built which in many places followed the railways' routes and went far beyond them (distribution). The *Major Traffic Street and Highway Plan*, accepted by the city council and the citizens in 1924, was intended to provide sufficient capacity to meet the demand for automobility. Although new roads were built and existing major roads were widened until the beginning of the Great Depression, the expansion of capacity did by no means keep up with the increasing volume of traffic (especially through more cars). The traffic problems were essentially seen as an expression of the far too little traffic on offer in the form of roads.

The Freeway Plans of the late 1930s continued to define mobility problems as technical tasks to be solved in order to improve the car transport system. The traffic problem then was perceived as the mixing of traffic from different directions at junctions. The intersection-free expressways or freeways were believed to be the technical solution to the traffic problem. After the Second World War and especially in the 1950s and 1960s, the traffic supply was increased by means of an extensive network of multi-lane freeways. These also had a much higher capacity (measured by traffic flow per car) than simple roads. Although freeways cover only 4% of the surface area of all roads and highways, they carry 40% of the traffic of the metropolis (Brodsly 1981: 9).

Initially, the transport offer seemed to be sufficient to meet the mobility needs of the population. However, as early as the end of the 1960s, congestion at peak traffic times indicated that the capacity limits of the new mode of transport had already been reached. Despite hundreds of freeway miles, the increased supply could not keep pace with the growing "automobile demand". The freeway system — again self-reinforcing — created its own demand over and above the increased demand caused by population growth (cf. above: Section 3.4.). Even as the freeway expansion slowed down in the early 1970s along with the *unintended byeffects* (smog, accident rates), which were now perceived more sensitively, the

---

[1] The controversial public debate on the 1920 parking ban deviates significantly from this pattern. By directly attempting to restrict the conditions of use of the car as a means of transport, the parking ban can be classified in the category of *negative* supply orientation.

skepticism in the population toward further freeway projects also increased.

Once again, decision-makers remained faithful to the supply-side orientation and looked for a mode of transport with a higher capacity to solve the mobility problems. They found it in rail-bound public transport. Since the end of the 1960s, a high-speed rail network has been promoted to meet the mobility needs and solve the traffic problems of the metropolis. In three (obligatory) referenda (1968, 1974, and 1976), citizens initially rejected the rapid transit plans. It was not until 1980 — after far-reaching planning and financial concessions — and again in 1990 that citizens agreed to an increase in VAT, among other things, to build a rapid transit network. As can be seen in Table 5.1, each defeat was followed by a rapid transit plan with an even greater range of services in the form of an extension of the route network (distribution).

The 1992 30-Year Transportation Plan published by the LACTC also showed a strong supply orientation. With a total network of 400 miles, rail transport was to be extended across the entire metropolis. In order to demonstrate the high capacity of this mode of transport, the Commission characteristically compared the potential performance of the railways with the capacity of a freeway lane which can handle 1,700 cars per hour (LACTC 1992: 18). Commuter trains replaced two lanes, light railways seven, and express and metro trains 14 freeway lanes per hour. One bus or carpool lane (High-Occupancy Vehicle Lane) can still carry the "Freeway Lane Equivalent" of five conventional lanes. Incidentally, in this comparison, the LACTC made an isolated reference to the capacity of railways, without asking the precarious question of whether the assumed maximum capacity utilization can ever be achieved even approximately.

The measures taken since the 1960s, known as "Transport System Management" (TSM), are less capital intensive but also maintain a supply

Table 5.1:   High-speed rail plans by year of referendum, route length, and estimated costs.

Year	Route length (miles)	Estimated costs (billion)
1968	89	$2.5
1974	145	$4.7
1976	232	$5.8
1990	400	$55.6

Source: Nelson (1983: 292 ff); Los Angeles County Transportation Commission (1992: 20.60).

orientation (Guiliano 1988). These include measures such as traffic monitoring and regulation, freeway access regulation (ramp entry control/ramp metering), and accident/breakdown management.[2] As financial resources for the construction of the freeway became scarcer, the metropolis attempted to improve the utilization of the existing network by means of TSM in order to avoid expensive extensions. Mobility problems are thus to be alleviated by optimizing transport technology to increase the capacity of the freeway.

In contrast to these supply-side strategies, since the 1980s, for the first time, measures have been deliberately taken to try to influence the demand side of transport. *Regulation XV*, for example, which has been outlined, aims to reduce traffic volumes by changing the behavior of individuals. The *Traffic Reduction and Improvement Program* (TRIP) can also be classified in this category since it is primarily intended to reduce individual demand for transport by car-pooling or job tickets. In contrast, *Proposition U* — although very hesitant — directly addresses mobility structures. More balanced land-use structures (commerce/housing) are to influence the development of transport. These more recent measures can certainly be seen as an expression of a conceptual rethink of transport policy.

According to Wachs, even long before the introduction of these mobility structure and demand-oriented measures, there was agreement among transport policy actors that specific mobility structures and behavior patterns were the main sources of transport (1990: 241). Nevertheless, there was also a consensus of values among the actors regarding the task of transport policy: To adapt the supply and capacity of the transport system to the growing demand for mobility instead of controlling mobility structures or behavior. The dominant transport policy supply orientation since the 1920s has been an expression of this consensus, which also

---

[2]These three measures were systemically integrated, starting with a 41-mile freeway test track as early as 1971. Sensors underneath the carriageway measure the traffic flow and transmit the data to a central computer which decides how many cars are granted access to the freeway, which is regulated by traffic lights. Unusual interruptions in the traffic flow — again detected by the electronic sensors — alert the traffic control center to breakdowns or accidents. In cooperation with the motorway police and a fleet of tow trucks, the congestion can be cleared relatively quickly. In addition, the monitoring center informs drivers about the traffic situation on the freeway network via radio and television stations. Today, 345 of the 730 freeway miles in Los Angeles, Orange, and Ventura Counties are electronically monitored (Endo/Janoyan 1992).

reflects the prevailing regulatory culture in the metropolis (cf. Section 6.3.1).

Nevertheless, the 30-year transport plan of the metropolis shows that the supply orientation of transport policy still prevails by far. Of the $108 billion already committed to projects, only $0.6 billion, or 0.5% of the budget, has been allocated to demand-driven measures. In contrast, more than $107 billion is to be spent on measures to increase the capacity of transport systems, albeit largely on public transport (LACTC 1992: 61).

5.2.2 Impact assessment of action patterns

The transport policy types of action (mobility structure, demand, and supply orientation) in turn form the background against which the dominant transport policies in Los Angeles are assessed in terms of their effects on mobility problems. Table 5.2 attempts — without claiming to be exhaustive — to classify transport policy strategies according to their depth of impact.[3] The deeper the measures are arranged, the more deeply they intervene in the conditions under which traffic is generated. Since there are considerable interactions between the complexes of measures, this model only has a heuristic function.

According to this model, the dominant supply-oriented transport policy in Los Angeles has only a low depth of effect. The hallmark of supply-oriented transport policy is to improve the conditions for the possibility of overcoming space.[4] In particular, the construction of the

---

[3]Prittwitz presents a step-by-step model for the environmental sector, according to which political action has different levels of effectiveness depending on the degree of problem-solving. Measures which start at the origin of the problems (e.g., population, structural, and technology policy) have a high depth of impact. Compensation, rehabilitation, or (burden) distribution policies, on the other hand, have a low depth of impact insofar as they deal with the consequences of problems, only combat the symptom, or postpone the problem (1990: 54 ff). The following considerations were inspired by Prittwitz's effect depth model.

[4]Supply-oriented measures can also aim at improving the enabling conditions for non-motorized modes of transport, for example, by building extensive cycle paths. In contrast to improving the supply of motorized modes of transport, these measures have a consolidating effect on mobility structures and are also very efficient in terms of land consumption and land use. They cannot be dealt with in this paper. Therefore, the supply-oriented measures are limited to motorized modes of transport.

Table 5.2: Transport policy strategies by depth of impact.

Low depth of impact
Offer orientation/ orientation toward motorized mode of transport
• *Traffic System Management (TSM)*: Traffic guidance systems, traffic monitoring, traffic rules, traffic signs
• *Strengthening individual transport*: Road and motorway construction
• *Strengthening public transport systems*: Establishment of bus and rail networks
Demand orientation (mobility behavior)
• Promotion of traffic avoidance
• Taxation of transport-intensive lifestyles, e.g., gasoline taxes
• Incentives for ecological transport modes
• Substitution of transport: telematics, telephone, fax, home office
Mobility Structure Orientation
• *Densification of activity locations*: Balanced use of space for housing, work, consumption, and leisure, subcentral activity centers
• Restrictions on access to certain areas
High depth of impact

freeway network in Los Angeles has shown that improving the supply of motorized modes of transport only initially has a traffic-reducing effect by distributing traffic over a wider range of services. However, an improved supply has a short-term (new) traffic-inducing effect on the mobility behavior of citizens: The need for more journeys is awakened, as these have become more attractive due to an improved supply.

On the other hand, policies that increase the supply of motorized modes of transport lead to a more distant mobility structure in the medium and long term, as citizens adapt to the new conditions and change their places of activity in the longer term: *Green* residential areas, more distant workplaces, consumer centers, and leisure facilities are chosen that change the structure of the area with their own dynamics. The possibility of increased spatial accessibility (mobility potential) transforms to a necessity (mobility compulsion) (cf. Owen 1959, 1966; Meyfahrt 1988; Ullrich 1988; Cerwenka 1982; Monheim/Monheim-Dandorfer 1990).

Public transport also has a low depth of impact in that it improves the possibilities of distance-intensive spatial conquest, as is the case in Los Angeles in particular. However, the denser and more compact the settlement structures are, the greater their effect. The transport system management (TSM) measures, which were increasingly introduced in the 1970s in an attempt to optimize existing transport services, also have a very low depth of impact. An example of this type of problem management is the Freeway Service Patrol Program which has been very popular in Los Angeles (Finnegan 1992). Breakdown and towing vehicles patrol selected sections of freeways to restore traffic flow as quickly as possible in the event of disruptions (accidents, breakdowns) and to reduce congestion.[5] Giuliano rates TSM programs positively because they are easy to implement and do not require large capital investments, and because they "complement the pervasive attitude among highway planners, policymakers, and users that traffic demand should be accommodated rather than controlled" (1988: 158).

In contrast, the demand-oriented measures introduced in Los Angeles in the 1980s (e.g., Regulation XV, TRIP) have a medium depth of effect. While supply-side policies do not attempt to control the volume of traffic themselves, demand-driven measures aim to reduce traffic through individual behavioral changes. It is true that changes in mobility behavior, such as car-pooling and transport mode change to the railways, do not bring activity locations closer to each other. However, positive and negative incentives (e.g., taxes) can indirectly influence mobility structures by making transport-intensive behavior more difficult and encouraging traffic-avoiding behavior. In the case of Los Angeles, however, the demand-oriented elements contradict the dominant supply-oriented measures. Since the latter usually generate additional traffic, it is to be expected that the demand-oriented elements, which aim to reduce traffic, will have little effect in this weighting.

It is difficult to assess the impact of the new communication technologies ("telematics") on transport. They offer the possibility to communicate and work over long distances without the need to travel. If telematics could be used to replace large traffic flows, it would be a highly effective instrument (cf., e.g., Nilles *et al.* 1976; Dover 1982). It is assumed, however, that telematics could also generate additional traffic in the medium and long term (cf. Bannon *et al.* 1982; Heinze/Kill 1987). For these reasons, telematics in Figure 20 has been grouped only at the medium level of impact.

Evaluations of the effects of telematics on traffic in California do not give a uniform picture. A study of the traffic effects of teleconferencing in

---

[5]The introduction of flexitime work is also of little effect as a traffic avoidance measure, as the volume of traffic itself is not reduced. However, traffic peaks can be smoothed out by distributing traffic in two ways.

Los Angeles concluded, for example, that it generates more traffic. It is true that the average distance to the various teleconference locations was 24% less than to the usual shared office. However, as more employees used the teleconferencing service, the total distance traveled was 29% greater than on an ordinary conference day (Mokhtarian 1988). On the other hand, "telecommuting", i.e., the total or partial replacement of commuting by telecommunications, seems to have some potential to reduce traffic significantly. Pendyala, for example, concludes that "on telecommuting days, the telecommuters made virtually no commute trips, reduced peak-period trip making by 60%, vehicle miles traveled by 80%, and freeway use by 40%" (1992: V). In the process, the traffic congestion for commuting to work did not lead to increases from other trips (Sampath *et al.* 1991). This suggests that *telecommuting* should be further reviewed as a traffic relief measure. According to a survey, 11% of workers in Los Angeles had the option of *telecommuting* in 1992. On average, however, they could only use this option for four days a month.

It has already been indicated that measures have a very high depth of impact if they allow for a densification of mobility structures. The spatial compression and mixing of activity locations, so that they are, at best, all accessible by bicycle and on foot ("city of short distances"), would be the optimal framework for efficient transport movement. Strategies oriented toward the structure of mobility thus address the conditions under which traffic is generated. In Los Angeles, there have been very few attempts to influence mobility structures (Proposition U). For suburban areas in particular, it is proposed, among other things, to establish subcenters in which both jobs and service centers are within walking distance (cf. Cervero 1988, 1986). Land-use planning (zoning) and tax incentives are repeatedly mentioned as possible instruments for this purpose (*ibid.*; Wachs 1990).[6]

---

[6] In addition to the control of mobility structures, it can also be assumed for the transport policy, in accordance with Volker von Prittwitz's step-by-step model of environmental action, that limitations in population growth have a very great depth of effect for certain areas. As the population of an area increases, so does its general need for mobility. Although some suburbs in California have introduced restrictions on immigration, this has been done only with the aim of maintaining the rural character and social homogeneity of their "bedroom communities" (cf., e.g., Müller 1981: 65) and not in order to enable mixed-use structures and avoid traffic. The fragmentation of the autonomous communities alone makes it almost impossible to envisage restricting the number of people moving into the city.

Finally, it should be reiterated that there is a high degree of interdependence between mobility structures, behavior, and modes of transport. Political action has a particularly high depth of impact when it influences the entire spectrum of mobility in a supplementary manner. Even good framework conditions (incentives for mixed use, cycle paths, etc.) can have little effect if the lifestyle of the population is not in line with the policy orientation.[7] By contrast, measures have a great impact if they are in harmony with the sociocultural orientations and lifestyle preferences of the population and can thus generate synergy effects (cf. below: Section 6.3; cf. also for the environmental sector Jänicke 1990).

Mobility structure-oriented measures only have a medium- and long-term effect. Changes in settlement structure and economic adaptation processes (e.g., through changes in land use regulations) are relatively rigid, similar to mobility-relevant values, attitudes, and behavioral routines, and can only be modified over a longer period of time. The latter are also difficult to control through political action.

---

[7]Authoritarian societies, for example, can pursue goals in a more coordinated and comprehensive manner, since they can also, for example, intervene in the behavior of the population to a far greater extent. The Kingdom of Singapore, for example, is known for its high degree of regulation of the mobility behavior of its population.

Chapter 6

Explaining the Development of the Transport Policy in Los Angeles

The L.A. Freeway is the cathedral of its time and place (...) Every time
we merge with traffic, we join our community in a wordless creed: belief
in individual freedom, in a technological liberation from place and cir-
cumstance, in a democracy of personal mobility. When we are stuck in
rush-hour traffic the freeway's greatest frustration is that it belies its
promise.

David Brodsly

The previous chapters have shown that both rail and automobile, as the
dominant modes of transport at their time, have had a similar effect
on the mobility patterns of Los Angeles. In addition, a predominantly
supply-orientated transport policy could be identified for all periods of
the metropolis's development. Political action reinforced the distance-
intensive mobility structures already established by the railways. The
metropolitan rapid transit plan, which is currently in the implementation
phase, will also favor the separation of activity locations rather than con-
tributing to their densification.

The following is an attempt to explain some elements of the development of the transport policy in Los Angeles. Special attention is given to
the interrelationships between modes of transport, mobility patterns, and
supply-side policy action. Hereinafter, the focus will first be on the
railway as a mode of transport. The decline of the tram and regional railways in Los Angeles will be explained by political–economic variables.

97

In addition to Bradford Snell's conspiracy theory (Section 6.1.1), the private sector organization of railway companies is examined as a determining factor in their decline. The mobility structure as an intervening variable is of particular importance.

The relatively autonomous municipalities scattered over a large area are the starting point for explaining the supply-oriented action patterns of transport policy in the metropolis. It is argued that supply-side policies are structurally more successful because of the competition of interests between the municipalities and their institutional blockade power.

Finally, sociocultural factors are used to examine the relatively permanent and fundamental subjective dimensions of (transport) political and social action in California. They should make the interrelationships between transport policy, modes of transport, and mobility patterns more plausible. The study begins by explaining why the (transport) political actors had little or no influence on the (planned) development of the city and metropolis and on transport (Section 6.3.1). Then, sociocultural factors are used to discuss the citizens' predominant preference for suburban housing (Section 6.3.2). Finally, the influence of the symbolic dimensions of modes of transport on individual behavior and transport policy action will be examined, which should make the irrational elements of mobility in Los Angeles clearer.

6.1   Politico-economic determinants of the decline of the railways

6.1.1   The conspiracy theory

The thesis formulated by Bradford Snell at hearings before the US Senate in 1974 on the decline of the railways and car ownership in America received a great deal of attention. After that, the urban railway companies fell victim to a *conspiracy* of the automobile industry and its suppliers. According to this theory, the automotive conglomerate General Motors (GM) in particular, together with co-conspirators from supplier industries, sought to destroy the efficient electric light railways and replace them with motor buses.

In addition to the production of cars and trucks, General Motors also dominated the production of buses and locomotives in America. According to the analysis of Snell (1973, 1974), the electric railways were an increasing competition for the automobile in urban transport. The more efficient the railways were, Snell pointed out, the fewer cars GM would sell. But selling cars was much more profitable for GM than selling buses and trains:

> The economics are obvious: one bus can eliminate 35 automobiles; one streetcar, subway or rail transit vehicle can supplant 50 passenger cars; (...) Due to the volume of units displaced, GM's gross revenues are 10 times greater if it sells cars rather than buses, and 25 to 35 times greater if it sells cars and trucks rather than train locomotives. The result was inevitable: a drive by GM to maximize profits by wrecking America's rail and bus systems (Snell 1974: 4).

In order to sell as many cars as possible, the conspirators purchased light rail vehicles throughout the country and replaced them with inefficient and uncomfortable diesel buses, which subsequently — as intended — lost passengers and went bankrupt in many places. Since citizens were now largely dependent on the car as practically the only motorized mode of transport, the conspirators had achieved their goal (*ibid.*: 5). According to Snell, the high operating costs of buses and their slow movement on congested roads encouraged the collapse of hundreds of public transport systems and the turning of passengers to cars. "In sum, the effect of General Motors' diversifications program was threefold: substitution of buses for passenger trains, streetcars and trolley buses: monopolization of bus production: and diversion of riders to automobiles" (Snell 1973: 27). Nowhere, according to Snell, was the ruinous impact of this motorization campaign more evident than in Los Angeles:

> Thirty-five years ago this was a beautiful city of lush palm trees, fragrant orange groves and ocean-enriched air. It was served by the world's largest railway network. In the late 1930's General Motors and allied highway interests acquired local transit companies, scrapped the pollution-free electric trains, tore down the power transmission lines, ripped up the tracks, and placed GM motor buses on already congested L.A. streets. The noisy, foul-smelling buses turned earlier patrons of the high-speed rail system away from public transit and, in effect, sold millions of private automobiles. Today, this city is an ecological wasteland (Snell 1974: 4).

In 1944, *American City Lines*, a subsidiary of *National City Lines* (NCL), bought the LARY tram network for $13 million. *General Motors, Standard Oil of California, Firestone Tire and Rubber Company, Phillips Petroleum Company*, and *Mack Truck* were among the shareholders of NCL. Within four months, the new management began replacing 19 of the 25 tram lines with diesel buses (St. Clair 1986: 61). National City Lines had to sell LARY two years later because it was accused by the Federal Court of violating *antitrust laws.*

Snell was severely criticized not only by General Motors but also by neutral observers because some facts were simply wrong or sloppily researched (General Motors 1974; cf. also: Adler 1991: 54 ff; Bottles 1987: 1 ff). However, part of his analysis still had verifiable explanatory content that explained the decline of the railways. It is true that from an economic point of view, motor buses were inferior not only to trams but also to electric (overhead line) buses as an urban mode of transport. In an economic aggregate analysis of the public transport modes of the motor-bus, trolleybus, and tram in US cities, St. Clair concluded that diesel-powered "motor buses were consistently the least economical transit vehicle during the period 1935 through 1950. Trolley coaches were significantly more profitable than motor buses. Streetcars were also more profitable than motor buses, although less profitable than trolley coaches" (1981: 599 f).

The economic superiority of electric buses over motor buses was also confirmed in other analyses (cf. Jones 1985: 52 ff). In addition, in the 1940s, there were documented cases in which NCL put motor buses into service though the mostly more profitable trolleybuses had already been ordered and work on the overhead line system had already begun (St. Clair 1986: 75 f; cf. *ibid.*: 56–80).[1]

These observations supported Snell's hypothesis, which assumed, among other things, that trams were replaced by unprofitable (and uncom-fortable) diesel buses, in order to let them subsequently perish ("conver-sion for destruction"). The underlying motive of GM and its suppliers to make more profits by selling cars rather than buses explained NCL's — otherwise irrational — motorbus orientation.

---

[1]The observed superiority of trolleybuses should not mean that there was no economic place for motor coaches in the local transport system. David St. Clair rather suggested that "indeed a mixed system would appear to have been optimal" (St. Clair 1986: 114).

In this respect, it seems clear that GM had some influence on the decline of public transport in the USA and in particular on the Los Angeles Railway (LARY). NCL's entry into the public transport market led to the dominance of a single diesel bus producer (GM) in the market and to the preemption of routes and markets which could have been served much more efficiently by trolleybuses. According to Jones, GM's dominance in the bus market...

> ...served to dampen competition in transit coach manufacturing and thus probably retarded innovations in both diesel and electric bus design *that might have reduced transit operating costs* (Hervh., S.B.). In the absence of significant competitive incentive, few improvements in diesel bus technology were made after the 1950s, although improvements of European origin established that the potential for advancement was not yet exhausted (Jones 1985: 63).

However, the conspiracy theory cannot explain why in Los Angeles the replacement of railway lines by buses began even before NCL became involved in local transport. For example, PE started to close down regional train lines and replace them with buses before GM was even active in the bus business. Snell also made a false observation in this connection, assuming that the railways were economically sound and ensured efficient operation before GM intervened (Snell 1974: 4). However, this was not the case for Los Angeles in particular, since LARY and PE were in an extremely poor economic conditions in the 1920s and 1930s. This was due not least to the sprawling mobility structures.

6.1.2 The private sector organization of transport companies

As private organizations, the railway companies were interested in transporting many citizens as far as possible. The dispersal of the population through investment in a huge rail network since the end of the 19th century was intended — in addition to the direct proceeds from real estate speculation — to lead to growing passenger numbers in the longer term and generate profits. Until the First World War, this calculation worked out for the railways, which had merged into two large companies. First, due to the distance-intensive mobility structure of the metropolis, an Angeleno used the railways much more frequently than a citizen of a city of comparable size. And second, rising population figures indicated a growing need for transport. This promised the companies profitable business for the future as well, due to the dominance of the electric railways in local transport, which was still unaffected at the time. However, the LARY and the PE unexpectedly faced competition from other motorized modes of transport in the period that followed.

From 1914 onward, the volume of freight transported by both companies declined rapidly (Fogelson 1967: 168 ff). While it was possible to make up for the absolute slump in passenger numbers from 1914–1917 with the population boom of the 1920s, the passenger volume per capita showed a steady downward trend at PE (Figure 6.1).

The sudden and vehement slump in rail passenger numbers could not (yet) have been triggered by the new private transport mode of the car. In Los Angeles County at that time, there were just 55,000 cars (1915) registered — too few to cause this shift (Brodsly 1981: 82). Instead, a new mode of transport, the *jitney*, had been on the streets of L.A. since 1913: private car owners offering pedestrians — in competition with the railways — their transport services by running without a timetable on the city's most busy major roads and carrying passengers on a five-cent basis (Nelson 1983: 269). By December 1914, there were already 1,800 licensed *jitneys* operating, earning the five-cent fares for 150,000 journeys a day, which had previously gone into the coffers of the railway companies (Richmond 1991: 34).

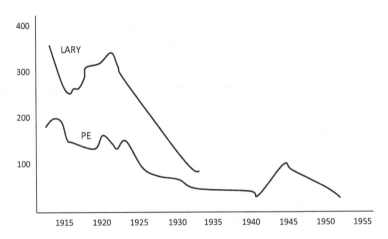

Figure 6.1:   Annual per capita journeys by rail (including buses).

Source: Based on Bottles (1987: 266).

The success of the *jitneys* was based on several advantages over the railways. On the one hand, they were more flexible: They did not follow a uniform route, did not rely on rails, had no timetable, and were numerous on many roads. On the other hand, they could work cheaply: Apart from a minimal license fee, they paid no other taxes or levies. This gave them an unfair advantage over the railway companies, which paid both considerable state and local taxes. In addition, the railways had to maintain the roads on which their tracks were located, which in turn benefited the jitneys — and later private transport. Since they did not have to pay VAT, the public authorities also lost income. After they had to pay taxes and duties as public transport companies in 1917, they quickly disappeared from the cityscape, as this had made their business unprofitable.

In addition to the *jitneys,* the railways also faced an adversary in regional transport in the metropolis at around the same time: Motorbus companies began operating between Los Angeles and the surrounding area, thus becoming the main competitor for passengers on regional trains. The loss of passengers was so dramatic on some parts of the PE that its president, Paul Shoup, said,

> Your Long Beach lines have fallen to the point where they (...) do not
> even pay the transportation expenses, that is the power to move the cars
> over the tracks (quoted after: Richmond 1991: 34).

To limit the ruinous competition with buses, since 1917, the railways had been buying up bus lines that competed with passengers on their own lines. As early as 1927, PE operated 32 bus lines with a network length of over 200 miles and carried over 820,000 passengers a month (Richmond 1991: 34 f). In 1939, 35% of PE's total passenger miles were covered by buses (General Motors 1974: 28). In this respect General Motors' answer to Snell's accusations is correct: "both the Pacific Electric and the Los Angeles Railway began to abandon streetcars before GM was even in the bus business and long before National City Lines or any affiliated companies were even organized" (General Motors 1974: 27).

However, the operation of buses was often unprofitable for railway companies.[2] This aggravated their poor financial situation, which was

---

[2] Jones emphasized, for example, that "(m)otorbus extensions of street railway lines were frequently deficit operations and almost invariably reduced a street railway's net income" (1985: 53). But, Jones continued, "(O)perating such services allowed street railways to

heavily burdened by the high capital costs of the extensive rail network. Railway companies had become heavily indebted in their efforts to make speculative profits for their real estate companies by building railway lines. By 1910, the debt burden of most railway companies already amounted to half of their total assets. The annual interest alone amounted to around $1.2 million for the LARY and $3 million for the PE (Fogelson 1967: 164). Both — like most rail operators in the US (Whitt/Yago 1985: 42 f) — had problems in acquiring sufficient investment capital.[3]

The competition from *jitneys* and motor coaches led to painful losses of income for the railway companies, which since 1914 had led to considerable losses of profits (see appendix). While the LARY was back in the black figures in 1920, the PE slipped into a deficit, which even exceeded the $2.5 million loss limit at the end of the First World War. Since 1930, the income of the PE could hardly cover the operating costs (Fogelson 1967: 184).

Although, unlike the PE, LARY was mostly in the black figures until the late 1920s, it only expanded its network by 24 miles between 1913 and 1925, while the city's population doubled during the same period. Instead of expanding their railway network, they opted for the less capital-intensive investment in motor bus companies compared to line extensions.[4] The renewal of old tracks alone in the 1920s came very close

---

preempt a latent competitor, satisfy demands for *service at a lesser capital cost than would have been entailed in rail extension* (emphasis by S.B.), court goodwill of city officials, and, at least in theory, build a traffic base that would generate future profits" (Jones 1985: 55).

[3] David St. Clair also pointed to overcapitalization and diluted share capital of railway operators, stressing that they themselves were "evidence of more fundamental regulatory failure" (1986: 96). The political institutions in Los Angeles had not been able to deal with these problems early enough. In 1913, for example, the California Railroad Commission rejected a LARY application for necessary rehabilitation (*ibid.*). Moreover, many years later, the Railroad Commission still lacked the powers to deal with problems such as overcapitalization and debt. For example, in 1938, according to the wording of a memorandum of the Commission, even in the case of the overindebted PE, "(t)he only suggestion (...) to persuade the parent company, Southern Pacific Railway, to voluntarily reorganize Pacific Electric, including the writing off of $43 million debt owed Southern Pacific, and completely refinance the remaining debt" (quoted after: *ibid.*). The other alternative would have been the bankruptcy of the PE.

[4] Dewees (1970) stressed that for the United States the operating costs of buses in 1929 were greater than those of trams. "But because of the fixed investment in rails and power

to the cost of completely new lines. Elsewhere in the United States, too, many companies at that time were considering replacing their railway lines with bus services when investments were due (Dewees 1970: 571).

As early as 1941, even before the National City Lines bought up LARY, the company considered abolishing many of its tram lines and setting up bus lines instead (Adler 1991: 56). Similarly, the PE in 1939 declared that "it had filed with CRC (California Railroad Commission, S.B.) an extensive rehabilitation program, that involved substituting buses for existing rail lines" (*ibid.*: 57). The Southern Pacific, the parent company of PE, was in any case said to have lost interest in passenger transport as early as the mid-1920s in order to concentrate on the much more lucrative freight transport (Hilton/Due 1960: 409).

While the conversions were still delayed due to the rationing of fuel and rubber for the Second World War, the Southern Pacific definitively declared in 1949 that it was no longer prepared to subsidize rail transport in Los Angeles any further (Adler 1991: 60). In the years that followed, the company replaced buses with rail lines as soon as it received the necessary permission from the California Public Utility Commission. In 1961, the city-owned Los Angeles Metropolitan Transit Authority, as the new owner of PE, discontinued passenger rail transport.

Under these economic conditions, at least for Los Angeles, the conspiratorial intervention of General Motors and its supplier industries *does not* seem to have been *decisive* in the decline of the electric railways. Rather, we must agree with Brodsly's assessment: "It required no conspiracy to destroy the electric railways; it would, however, have required a conspiracy to save them" (1981: 95).

However, the conditions for a conspiracy of political actors to save the city's railways were poor in the 1920s and even decades later. Local public transport in Los Angeles was "viewed as a business that was expected (...) to pay its own way and earn profits for its shareholders" (St. Clair 1986: 100; cf. below: Section 6.3).

From then on, rail transport in Los Angeles was only possible with substantial financial support. Ironically, the railways themselves had contributed to this. The horizontal settlement patterns and the expected high demand for transport did not lead to the hoped-for profits but turned

---

distribution, if service was to be extended to a new corridor, there would be some minimum traffic level below which the higher operating costs of a bus would be offset by the saving in ground investment" (Dewees 1970: 570).

against the rail companies. Instead, the urban *sprawl* of the metropolis provided a *good starting point* for the automobile, which became affordable for broad sections of the population in the 1920s.[5] In view of the low population density, the railway could not compete with the much greater spatial flexibility of this motorized means of individual transport. The car now made it possible to settle down between the previously inaccessible corridors of the railways, considerably shortened travel times between the suburbs, and simplified the realization of a low-density lifestyle (cf. Section 6.3.2).

Together with road construction, the new means of transport, the automobile, developed a momentum of its own that changed the mobility structure of the metropolis. This new environment, characterized by even more extreme urban sprawl and increasing decentralization of jobs, led to a thinning of traffic flows in the region. As a result, rail-bound public transport became uneconomic and, compared to the car, unattractive for the individual. Motor buses, which did not rely on fixed routes, were able to adapt better to the dynamically changing mobility structures since the 1920s and also benefited from the road and freeway construction of the following period. In this respect, the decline of the railways should not be interpreted as a simple consequence of competition from the car as a means of local transport. Rather, the changes in the mobility structure and mobility behavior, which were considerably accelerated by the car later, are decisive. These were able to develop in such a dynamic way, not least because of the considerable subsidization of car traffic.[6]

---

[5]In addition, the Los Angeles electric railways supported — albeit involuntarily — the competing modes of transport, bus and car, because they had to co-finance road construction. As described above, the city's concession conditions required the railways to pave the area adjacent to the tracks.

[6]This was also the view of Dewees about the US, "(W)hile the automobile was important, its major effect was to change the environment in which the street railways operated so that their cost rose rapidly relative to other means of travel" (1970: 579). With regard to the construction of transport routes after the Second World War, Jones said, "(T)he speed of travel and locational flexibility freeways permitted was an intimate part of a complex process of relocation and reorganization that transformed urban activity patterns in the postwar period. Changes in the geographic and temporal pattern of travel would deprive transit of the traffic density and traffic balance needed to sustain profitability" (1985: 72).

6.2 Institutions, location interests, and supply-oriented policy

The mobility structures of the metropolis form the starting point to explain the supply-oriented action patterns of transport policy. The Los Angeles Region (CMSA) covers a comparatively large area of around 88,000 sq. km (San Francisco 19,000 sq. km; New York 18,000 sq. km; cf. Table 1.1). The inhabitants of the 200 or so autonomous communities located in this area needed good transport connections to reach their — geographically widely separated — places of activity. Transport systems improved the accessibility and thus the attractiveness of these areas. Since each municipality tried to increase its attractiveness, there was competition between the various municipalities, especially for investment in transport infrastructure (cf. Adler 1986).

In the following, it is argued that, because of these locational competitions, politics between the municipalities in the metropolis was highly conflictual. Due to the institutional blockade power of the independent cities, only those policies in the metropolis can be successful which distribute the benefits spatially as evenly as possible among the municipalities. The supply-orientated characteristic of transport policy fulfills this requirement.

The planning sovereignty of the municipalities made it difficult to coordinate planning in the Greater Los Angeles area. Strong policy coordination in turn would be a prerequisite for changing the distance-intensive mobility structures. In contrast, each individual municipality exercised the land-use authority in its own area, so the metropolis was characterized by a variety of different planning policies aimed at maximizing the respective benefits of its municipality. "In city planning operations, the dominant formal goal is maximizing benefits and minimizing costs. (...) Each local government strives for a 'desirable' pattern of development" (Logan/Molotch 1987: 187). In order to maximize individual benefits, each city government estimated which future land-use regulations would bring it the greatest benefits. Two strategies can be distinguished, which were usually combined. On the one hand, municipalities could limit the number of new development areas or restrict them to highly valued areas. In both cases, this meant that only wealthier citizens could afford to settle down. Apart from fiscal aspects, the aim was of course to maintain a homogeneous (white) population. On the other hand, the other strategy was to use incentive schemes (offering cheap land, subsidies) to attract companies in order to use the income from business tax to cover the increased expenditure on municipal services.[7]

The economic differences between the communities alone held considerable potential for conflict, which made coordinated action difficult. As outlined earlier, the poorest classes lived in the central areas of L.A., while the better-off citizens lived in the suburbs of the metropolis. Logan and Molotch pointed out, among other things, that the land-use competences of the communities guaranteed and reinforced these inequalities. Since the poor communities had to cover municipal services, such as schools and police, from their own meager tax revenues, a vicious circle of poverty was inevitable (1987: 195 ff). The economic heterogeneity of the communities also reflected their different interests. While the poorer communities in the center, for example, considered better bus transport to be necessary for the metropolis, richer suburban communities did not see this investment as a priority.

These factors explain the supply-side orientation of the transport policy in Los Angeles. Since the political arena of the metropolis had a very conflictual character, public measures that minimized the potential for conflict appeared more promising. According to the thesis of Lowi (1964; 1972), the type of policy determines the level of conflict in the political process ("policy determines politics"). Distributive measures, such as land allocation, subsidies, and patents, then generated the least conflict in an arena if the benefits were distributed among a large number of addressees. Thus, if the actors expected everyone to benefit from a measure in roughly the same way, there would usually be little resistance to it.[8] According to this concept, supply-side transport policy measures in

---

[7]This strategy took on an important role after the freezing of the real estate tax. For example, "Proposition 13" following the tax revolt of 1978 limited the rate of assessment of the municipal real estate tax to 1% of the market value of the base year 1975/76, thus reducing not only an important source of income but also the scope for action by the municipalities (Billeibeck 1989: 68 ff). In discussions, for example, leaders of the Southern California Association of Government (SCAG) complained that local authorities now "use land use planning excessively to attract industries", as this was one of the few means of covering the costs of municipal services (according to Ralph Cipriano (Principal Planner), interview on 19 November 1992, Los Angeles).

[8]It is true that, in the long run, all policies are redistributive since goods are finite and subsidies and the like have to be financed by tax revenues. But if one assumes, as Lowi

Los Angeles (in the sense of political enforceability) could be successful because they were *distributive* in nature (in Lowi's understanding). Especially in the metropolitan area of Los Angeles, where the municipalities had the institutional competence to block policies affecting them, conflictual policies were usually doomed to failure.

The dominant transport policies can now be examined in detail based on this concept. If one compares, for example, the financing of the "General Road Transport Plan of 1924" with the metropolitan rapid transit plan presented the following year, the automotive-oriented strategy showed considerably less potential for conflict. The construction of a rapid transit railway was valued at $133 million and was to be financed by bonds, property tax, and fare increases (Wachs 1984: 308). In addition, the plan was characterized by a strong radial orientation and thus favored the center of the metropolis unilaterally (Foster 1976: 469 ff). On the other hand, the wider network of the road transport plan seemed to benefit, to a far greater extent, the exploding communities in the Los Angeles region.

In order to keep the level of conflict low, it was perhaps even more important that the city council was very cautious in its financial demands and only ever demanded small sacrifices from the citizens. Thus, in 1924, the city council started with a modest $5 million dollar bond issue, followed by a temporary increase in property tax only in 1926, both of which were accepted by the citizens in referenda. In contrast, the rapid transit plan — much more obviously — required much greater financial sacrifices in the short term, as the total cost was incurred very quickly. This gave them a clearer character of redistribution than the perceived costs of the road transport plan. In this respect, the rapid transit plan can clearly be placed in the category of *redistributive policy:*

> The nature of the redistributive issue is not determined by the outcome of a battle over how redistributive policy is going to be. Expectation about how it threatens to be, are determinative (Lowi 1964: 691).

The rapid transit plan was not very attractive to citizens because at first sight it promised much higher costs and required more sacrifice (higher fares) than the road transport plan.

---

did, that "politics run in the short run, and in the short run certain kinds of government decisions can be made without regard to limited resources" (1964: 690), then the distributive category makes sense.

The *distributive* character of the transport policy measures in Lowi's sense was most evident in the construction of the freeway network since the end of the 1940s. The federal government subsidized 90% of the freeways. Since the planned freeway system covered the entire metropolis with a dense network, every municipality could expect to benefit from the construction. At the same time, the costs were financed by federal subsidies and by the increase in Californian mineral oil tax and therefore offered little cause for conflict. The latter type of financing also had a conflict-minimizing effect, as the costs were subjectively perceived as low ("seven cents/gallon"). Furthermore, they appeared to be *fair*, as they applied to all motorists. In fact, there were hardly any conflicts between the communities about the freeways and their financing. In contrast, Nelson reported, "Argument was mainly over who should get their freeways first" (1983: 281).

The thesis that the respective policy determines the level of conflict in the political arena is also confirmed by the rejection of rapid transit plans since the late 1960s and their acceptance by the population in 1980 and 1990. In order to build a rail-bound public transport system in the metropolis, the responsible authority, the Southern California Rapid Transit District (SCRTD), had to minimize the conflicts arising from the competition between locations and different interests. Characteristically, this was again attempted with policies that had a distributive effect. The planned measures were all mainly aimed at providing the region with comprehensive services through railways, buses on freeways, and feeder lines (cf. Chapter 4). However, the population rejected the first plan of 1968 by a majority of 55 to 45%. After analyzing the votes, the SCRTD concluded that the service was inadequate for the area (Stipak 1973) and from then on made more extensive and considerably more elaborate proposals.

But citizens also rejected the plans in the referenda of 1974 and 1976. The opposition was made up of taxpayers from the surrounding area, on the one hand, and municipalities that were not well served by the planned network, on the other hand, but were forced to pay for it — in their opinion (Marcuse 1975: 29). Indeed, according to Marcuse's analysis, "the ridership benefits of the system would have gone by and large, to the white, middle and upper class, to white collar employees and executives and Professionals commuting to work downtown from suburban residences" (1975: 18). In contrast, the poor, blue-collar workers, homeowners, and businessmen who lived far away from the stations would have to bear the costs in an unreasonable way due to the financing by the VAT (*ibid.*).

To win the approval of these citizens for the 1980 plan, the newly formed Los Angeles County Transportation Commission (LACTC) not only significantly expanded the proposed rapid transit network but also made extensive concessions to the communities and socioeconomically disadvantaged groups that benefited little from the rapid transit network. For example, only 50% of the proceeds from the VAT increase were to be used for railway construction. A quarter of the money was to be used to finance regional bus lines and traffic management measures, while the remaining 25% was to be returned to the municipalities. To win over the socially disadvantaged existing users of the bus lines to the plan, it was promised that the revenues from VAT would be used to substantially reduce bus fares for three years (Richmond 1991: 145 ff; Adler 1987: 331 f). This broad distribution of benefits ensured the approval of a large number of heterogeneous groups in the metropolis, so in 1980 and 1990, a majority of citizens supported the increase in VAT by half a percentage point each.

Beyond the explanation of supply-oriented transport policy, the perspective of the (transport) policy process as a competition for locational advantages leaves room for another consideration. As discussed earlier, the municipalities that were institutionally favored defended their respective locational interests. From this point of view, investments in transport modes (road or rail) increased the locational advantages of a region by increasing its *reception capacity*, i.e., the number of workers and consumers who can reach a location, and could subsequently attract capital investment. Since investments in both road and rail networks in principle improved locational conditions, it was reasonable to conclude that the lines of interest and conflict within the metropolis ran less *between* transport systems (car vs. rail) than between municipalities, which more or less benefited from the respective routing of the transport investments. This consideration cannot be further explored in the context of this work. However, other studies indicate that this assumption is correct (Adler 1986; Foster 1976; Wachs 1984).[9]

---

[9]Sy Adler interpreted Los Angeles' transport policy development from the competition of interests between the center (CBD) and the business centers of the surrounding area (Adler 1986, 1987, 1991). In the 1920s, for example, political and economic actors in the center of L.A. pursued the strategy of "building a metropolitan scale *downtown/radial* (emphasis, S B.)

6.3 Sociocultural factors

In Section 6.1, it was shown that political and economic factors had a major impact on the decline of the Los Angles railways. Mobility structures were discussed as important intervening variables which had a decisive influence on the economic success of transport modes. Subsequently, it was to be shown that the locational interests of the communities, which had been strengthened by institutional fragmentation, could explain the supply-oriented orientation of transport policy. Finally, sociocultural factors are used to focus on the underlying value basis of (transport) political and social action (or inaction).

With America's political–cultural orientations and traditions, the first step is to discuss the reluctance of public decision-makers to intervene in urban and transport development. Particularly during the phases of high population growth from the turn of the century to the 1920, mobility patterns in Los Angeles were able to develop practically free of state regulation (Section 6.3.1). This is followed by an examination of the preference for the "low-density" lifestyle, which is so dominant for Los Angeles in particular and which represents an essential element of California's mobility structure (Section 6.3.2). Finally, the last section discusses the symbolic dimensions of the rail and automobile modes of transport, which are closely intertwined with sociocultural factors (Section 6.3.3). This is intended above all to make the hidden irrational elements of social action associated with cars and railways more plausible.

6.3.1 Political culture: Privatism and urban development

Many see in the institutional structure of Southern California, especially in the fragmentation of planning and in the plebiscitary elements of state

---

rail rapid transit system (...) designed to facilitate travel between centre and periphery, rather than within or between peripheral areas" (Adler 1986: 322). Rail rapid transit systems were more advantageous for the center than roads and freeways, as railways have higher transport capacities and the radial orientation of the network would stabilize the economic supremacy of the CBD. On the other hand, the growing number of business districts in the surrounding area — which already felt threatened by the dominance of the CBD — saw greater advantages in a road and freeway system, as this improved the connections between the business centers and their respective catchment areas (*ibid.*).

legislation, value realizations of individualist–privatist elements of political culture (Billerbeck 1989). James Q. Wilson notes, for example, that "(t)he well known institutional features of California's political system — weak parties, the extensive use of the referendum to decide policy issues, nonpartisanship — were perfectly matched to the political mentality that was nurtured in Southern California (...) The political institutions and economic character of Southern California reinforced the lifestyle and gave it expressions" (1967: 40, 41).

The liberal tradition of North America, which — strongly influenced by John Locke — has resulted in a very high value being placed on individualism, equality, freedom, and the protection of property is hardly ever disputed in the literature (Vorländer 1992: 303). This homogeneous canon of social values and political attitudes, which can be assigned to a liberal context of origin and tradition, has remained almost unchanged not only in the last forty years but also "for the period before that, for which there is virtually no systematic data collection, the persistence of liberal attitudes is assumed, mostly from historical studies" (*ibid.*: 304).[10]

The limited influence of political actors on urban development and transport planning is an expression of the individualistic–privatist elements of political culture. Referring to Sam Bass Warner, Heidenheimer *et al.* declared "that it has been private institutions and individuals that have been responsible for guaranteeing the productivity and social order in American cities" (1975: 104). The physical form of American cities is a visible example of the consequences of *privatism*:

> The townscape is determined by what Mumford has labeled the 'speculative ground plan', which treats urban land as a commodity whose worth is expressed strictly in terms of its market value (*ebd.*: 105).

---

[10]Of course, America's political culture cannot be traced back to liberalism alone. However, as Bellah *et al.* (1985) also stated, the liberal orientation, which is composed of puritan and republican contexts of justification and tradition, is dominant, which gives very high priority to the individual and individual development (cf. also Vorländer 1992: 304 ff). The Declaration of Independence, for example, also began with Locke's liberal elements of theory in natural law: All human beings are created equal and endowed by their Creator with inalienable rights, which include life, freedom, and the pursuit of happiness. Governments, it goes on to say, are appointed to secure these rights (cf. Zöller 1992: 281).

The individualistic *bias* in the American legal system prevented state decision-makers from intervening in existing structures when these were privately owned (cf *ibid.*). Thus, at the end of the 19th century, it was out of the question for city administrations to become active as operators of trams themselves. Fogelson stressed that "municipal authorities lacked (...) even the inclination to build and operate these facilities" (1967: 40). Rather, the city council, as the main political actor, did not usually stand in the way of private rail operators that wanted to invest. Usually, they granted them the desired concessions for transport routes.

Investment decisions by private transport and land development companies as well as population growth were important for the development of Los Angeles. In Chapter 2, it was shown that the growth of the metropolis only began with the arrival of the intercontinental railways in 1876. At the same time, the electric trams and regional railways developed into important means of transport. The coincidence of major migratory movements to Los Angeles and the availability of motorized transport made possible the widespread urban sprawl of the population, which was further intensified by the entrepreneurial intertwining of passenger transport and real estate speculation.

As private initiative was given broad scope, its effects were far more dominant than in Europe, for example. In Los Angeles, the profit motives of transport companies and real estate speculators were combined with the lifestyle preferences of growing population groups while at the same time restraining state regulation. In comparison to Germany, where city governments systematically controlled and planned the spatial location and appearance of new buildings as well as the maintenance and restoration of old buildings at an early stage, Los Angeles was able to develop for a long time practically without regulatory intervention by the government.

In contrast to Europe, there was a complete lack of planning regulations until the 1920s, which determined the spatial location of new buildings, their form (single-family houses/multi-family houses), and the way in which land was used (separation of use/mixing of uses). Urban and regional planning in Los Angeles, like the takeover of utilities by municipalities, was a child of the progressive movement at the beginning of the 20th century.[11] Progressives held the view that under the previous

---

[11] The progressive movement was a response to political and economic corruption, which was commonplace in California, in particular, to an extent unimaginable today. Huge fortunes were made in the big cities with building scandals and land speculation (Billerbeck

institutional arrangements, private developers made a profit and burdened communities with damage in the form of poor roads and premature land development, which led to congestion and increased community expenditure (Fogelson 1967: 247).

In 1915, the *Progressives* formed a voluntary association with the *City Planning Association*, whose members were to advise the city on planning issues. Five years later, the city council set up an urban planning commission, but until 1925, it had very little financial resources or authority. It was not until the adoption of the City Charter of Los Angeles, a separate city constitution, that the commission was given more money and authority. But even then, it had little influence on the spatial development of the city (Foster 1976: 480 ff). Thus, the Planning Commission with its subsequent regulations did not in fact change the practice of land development. Instead, the planners considered the horizontal structure of the metropolis, which had already been largely established in the 1920s, to be desirable. Their attitudes regarding urban form and land use corresponded to the residential preferences of the citizens. Many of them had already been able to turn these preferences into reality in the 1920s:

> From their conception of congested eastern and midwestern metropolises, the planners assumed that the great city was no longer the most pleasant place for living or the most efficient location for working. They proposed as an alternative residential dispersal and business decentralization (Fogelson 1967: 250).

The Planning Commission supported the trend of land-use separation. The single-family houses, which were also clearly preferred by the planners, were to be carefully protected and therefore separated

---

1987: 46). At their peak, the progressives managed to have the water and electricity supply as well as the harbor bought and operated by the municipality. At that time (around 1910), however, there was never any serious discussion about also transferring the railway companies into municipal ownership, as this would have required immense amounts of capital (Bottles 1987: 263). Instead, they established a number of authorities to regulate the utilities so that they would never again make a profit at the expense of the citizens. In 1909, the city council established the Board of Public Utilities. In addition to overseeing the gas, electricity, and telephone companies in the city, it also supervised the transport companies. Its main tasks were to set the fares of the LARY and PE, and to advise the City Council on the award of concessions for the renewal and extension of the railway networks.

from trade and industry as far as possible. Multi-family houses should act as a buffer between them. According to the planners' ideas, trade should be concentrated in the center and along the main roads and production should be limited to industrial sectors. "In sum, zoning reinforced land — use segregation in Los Angeles by investing the patterns imposed by private enterprise with governmental sanctions" (Fogelson 1967: 255).

Coordinated planning arrangements for the greater Los Angeles area were then, as now, hardly possible. Even at that time, planning competence was already the responsibility of the many independent municipalities in the metropolis. In 1923, a regional planning commission was established in Los Angeles County (as the first metropolis in the United States). However, it only had an advisory function and sought to mediate in problems and tasks that crossed administrative boundaries. Especially during economic downturns, it became apparent that the county's municipalities were not willing to relinquish control of land-use planning. Although the regional planning authority was praised for its efforts to formulate long-term development plans and to consult with all the authorities concerned, it was not clear whether this was the case. "Unfortunately," according to Foster, "even in this reportedly 'ideal' working relationship between planners and local government bodies, public officials in Los Angeles followed planning recommendations only when they conveniently served short-termed, obvious needs" (1981: 138 f). Instead of cooperating with each other, land-use planning was used as an instrument for asserting locational advantages against each other (cf. above: Section 6.2).

The individualist–privatist elements of political culture also predominated in the transport policy in the narrower sense. For example, shortly after the First World War, one of the first and most prominent transport engineers in the US expressed the view that passenger transport was subordinated to the individual interests of transport users: "(The) driving force, the motive that has negotiated franchises, consolidations, and mergers has been for profit only" (Wilcox, quoted after: Whitt/Yago 1985: 39). State intervention had mostly been limited to arbitrating the dominant interests of particular industrial groups. Even transport planning — after slowly beginning to take off in the second decade of the 20th century — was spatially very limited, short term, and ad hoc; "it commonly has not addressed the larger issue of the overall relation between transportation and city development" (Whitt/Yago 1985: 39).

One can only speak of traffic planning in Los Angeles in the 1920s. Construction measures had to be initiated beforehand by affected landowners by petitioning the city council. These were then examined by municipal civil engineers on a *case-by-case* basis (see above). Although the city had already been spending modest sums of money on road construction since 1909, it was not always possible to find a solution. But it was only when traffic in the metropolis became increasingly chaotic that a general road traffic plan was sought in the 1920s to get the problems under control.

Even today, the individualism of the communities prevents coordinated planning in the metropolis. The *Southern California Association of Governments* (SCAG) is the declared planning authority of the six-county metropolitan area. SCAG, together with the counties, cities, other agencies, and stakeholders, is responsible for the development and adoption of growth management and regional transportation plans (see, e.g., SCAG 1989a, 1989b). The formal and factual influence of SCAG on the regional and transport development of the metropolis is, however, according to its own statements, extremely modest since the competencies still lie with the numerous autonomous municipalities. Equipped with few competencies and financial resources, SCAG's main task is to coordinate the numerous actors horizontally (e.g., by means of persuasion) (Shaw/Simon 1982: 14 ff).[12]

6.3.2 The preference for suburban housing

We have described the tendency of the population to live in suburbs, which has been noticeable since the turn to the 20th century, as an important element of the mobility structure of the metropolis. However, no explanation has yet been provided for the citizens' preference for this "low-density" lifestyle,[13] which is essentially responsible for the

---

[12]These observations were confirmed in interviews with SCAG managers. For example, SCAG planner Ralph Cipriano reported considerable problems in bringing together the different orientations and interests of the municipalities. For example, the 1989 transport plan drawn up by the authority over many years was rejected by the municipalities and had to be rewritten.

[13]Since "lifestyle" has become a dazzling term in the sociological discussion, it will be briefly defined in the following. The lowest common denominator is understood to be a "typical unmistakable structure of behavior patterns visible in everyday life", whereby attitudes and the objective prerequisites for shaping life are often also included in this term (Hradil 1992: 28; for a discussion of different lifestyle concepts, cf. *ibid.*) A "low-density"

Angelenos' distance-intensive mobility behavior. Sociocultural variables are now to be examined as explanatory factors for this lifestyle. To this end, the previous section will be continued and, in turn, findings from North American cultural research will be used to explain the lifestyle in Los Angeles. Finally, it will be examined whether and to what extent the lifestyle has changed in the meantime.

Utilitarian individualism and puritanical rural ethics are repeatedly mentioned in literature as important roots of American culture (cf., e.g., Bellah *et al.* 1985; Kamphausen 1992; Zöller 1992; Muller 1981: 20 ff). Bellah and others, following Toqueville, underlined the importance of the individualistic orientations which constitute the "habits of the heart"[14]:

> Individualism is a calm and considered feeling which disposes each citizen to isolate himself from the mass of his fellows and withdraw into the circle of family and friends; with this little society formed to his taste, he gladly leaves the greater society to look after itself (...) Such folk owe no man anything and hardly expect anything from anybody. They form the habit of thinking of themselves in isolation and imagine that their whole destiny is in their hands (Bellah *et al.* 1985: 37).

Together with the puritanical–agricultural lifestyle of the 19th century, mythical ideas developed around the independent farmer:

> Certainly, powerful American myths have been built around the self-reliant, but righteous, individual whose social base is the life of the small farmer or independent craftsmen and whose spirits are the idealized ethos of the township (*ibid.*: 40).

The puritanical rural ideal, whose philosophical roots "lay in the Jeffersonian perception of democracy, particularly its interpretation of the agrarian doctrine of the eighteenth century Enlightenment" (Muller 1981: 21), has been in a certain tension with the rural exodus that began in the middle of the 19th century. Industrialization resulted in urbanization as an

---

lifestyle is simply understood here as the preference for living in low-density areas, especially in suburbs, and the associated (mobility) behavior patterns.

[14]Following Toqueville's use of the term, Bellah *et al.* (1985) call their influential book "Habits of the Heart. Individualism and Commitment in American Life". By habits of the heart, they mean "notions, opinions and ideas that 'shape mental habits'; and 'the sum of moral and intellectual dispositions of men in society'" (1985: 37).

unwanted by-product. The problem arose of combining the ideas of living in a rural idyll with living in cities, as mass production was only possible through the concentration of production facilities.

The preference for suburban living stems from this tension. Living in small suburbs brought one closer to the rural ideal while at the same time maintaining the unwanted but inevitable proximity to cities where one had to work. Anselm Strauss referred to the suburbs in this context as "(t)he Union of Urbanity and Rurality" (1976: 230). The suburban lifestyle can therefore be described as a value realization of individualistic and rural–puritanical orientations.

But why was the suburban lifestyle in Los Angeles much more dominant and pronounced than in other metropolises in the US? In 1930, none of the major cities in the US had such a low settlement density and high numbers of single-family homes as Los Angeles (cf. above; Fogelson 1967: 142 ff). Two factors played a decisive role here:

On the one hand, in contrast to the cities of the Northeast and East Coast, which mainly had immigrants from Europe, in the first half of the 20th century, a vast majority of newcomers to L.A. came from the *rural areas* of the Midwest of the US. Census data, for example, show that 1/4 to 1/3 of L.A.'s total population was born in the Midwest between 1890 and 1930 (Fogelson 1967: 79 ff). This trend of migration from the Midwest to Southern California was even evident in the years 1935–1940. As a result, the population of L.A. was characterized by a puritanical–agricultural ethos to a far greater extent than other metropolises.

On the other hand, Los Angeles also offered much better opportunities to realize the preference for horizontal living. While the cities on the East Coast matured into metropolises and developed a dense urban form even before the advent of trams and regional trains, Los Angeles only became a metropolis in the age of motorized transport. The largest network of trams and regional trains in the world enabled the population to expand geographically at a very early stage, thus enabling them to realize their preferences for suburban living. The already established urbanized structure provided a good starting point for the car and spread much faster in Los Angeles than elsewhere in the US. With the car, the Angelenos were then able to realize their ideas of low-density living even better.

While the cities of the east coast and northwest of the US were more European in character — due to the predominant origin of their population — the Angelenos sought to develop a new type of city. The rural newcomers from Southern California had an aversion to *the city of*

European color, which they considered immoral and inhumane. In Los Angeles, something else was to be created:

> Their vision was epitomized by the residential suburb — spacious, afflu-
> ent, clean, decent, permanent, predictable and homogenous — and vio-
> lated by the great city — congested, impoverished, filthy, immoral,
> transient, uncertain and heterogenous. The late nineteenth- and early
> twentieth-century metropolis, as the newcomers in Los Angeles per-
> ceived it, was the receptacle for all American evils and of all American
> sins. It contradicted their long cherished notions about the proper envi-
> ronment and compelled them to retreat to outskirts uncontaminated by
> urban vices and conducive to rural virtues (Fogelson 1967:145).

In the small, homogenous suburbs of Los Angeles, urban problems should not arise in the first place. This was to be achieved primarily through low settlement densities, a homogeneous population structure in the community,[15] and purely residential areas. To achieve this, the individual also accepted long distances to work. "The separation between the worlds of work and of family and community is often expressed and realised by the daily commute between factory or office and residential neighbourhood" (Bellah *et al.* 1985: 178). The suburban residential neighborhoods represented enclaves to which citizens fled from the city, which was characterized by crime, narrowness, dirt, and heterogeneity.

As an example of these attitudes, Bellah *et al.* cited a teacher living in a middle-class suburb of Los Angeles:

> When we lived in the city, we felt hemmed in. (...) I don't want drunk
> drivers driving like crazy down my street. It was just too crowded, too
> noisy. What's important to us is the sky, the quiet, and the space (...)
> A good community is when you have a complete mixture — enough
> shopping to take care of your needs, but not large shopping centers to
> bring people in from outside the community. I would want to see our
> community develop as if it were an island (1985: 180 f).

Since growth can shift the character of the community toward the hated "city", it poses a threat to their enclaves in the eyes of the

---

[15]As described, Los Angeles had largely achieved population homogeneity in the communities. The metropolis was characterized by sharp spatial dividing lines that ran between socioeconomic and ethnic groups.

inhabitants. Particularly in the suburbs of Southern California, there was therefore great resistance to further population growth and the designation of industrial projects in the municipality (Logan/Molotch 1987: 209 ff). Many citizens "were fighting desperately to slow down the development of their suburban community — to 'keep it from becoming like Los Angeles'" (Bellah *et al.* 1985: 180). The NIMBY (Not In My Backyard) movements (Dear 1992; Fisher 1993), in which citizens opposed public projects and "anti-growth regulations" were expressions of resistance. In the mid-1970s, the Growth Limitation Act of Petaluma, a suburb of San Francisco, caused a sensation: By means of a zoning ordinance, the community imposed a quota on the construction of new housing "in order to 'shape orderly growth' and safeguard the 'small-town character' of the community" (Muller 1981: 99).

Despite great efforts, the Angelenos could not prevent the change in the rural character of their suburban communities. While at the turn to the 20th century only the upper class and upper-middle class had reservations about living in the suburbs, in the 1920s and especially after the Second World War, broad sections of the population were able to participate in this lifestyle. Characteristically, "middle-class suburbs" and "working-class suburbs" developed in line with socioeconomic differences (Berger 1969). Sustained population growth now also led to a moderate densifica-tion of suburban areas. Since the 1970s, a majority of Americans had been living in the suburbs and — according to surveys — also wanted to live there (Müller 1981; Althshuler 1979: 377 ff). In addition, industrial projects could not be prevented in many places in what were once purely *bedroom communities*. These changes were already called "The Urbanization of the Suburbs" (Masotti/Hadden 1973) at the beginning of the 1970s.

The growing urbanization of the suburbs led to "Trouble in Paradise" (Baldassare 1986). Residents' satisfaction with their communities was declining: "Residents reacted negatively to the suburban transformation partly because their community ideals and expectations clash with the new realities of the disurbs" (Baldassare 1991: 208).[16] Both the traffic and environmental problems and the social problems no longer stopped at the former suburban enclaves. Flight from the problems of *the city* could only succeed as long as only a minority *fled*. The more the people moved to the

---

[16]The term "disurb" refers to bedroom communities or commuter suburbs which differ from urban regions mainly in their low population density and separation of use (Baldassare 1991: 209).

suburbs, the more they shifted the problems and the more the negative side effects from which they fled became apparent ("when everyone stands on their feet, no one can see better").

6.3.3 Symbols and mobility

The character of mobility in general and modes of transport in particular for society and politics in Los Angeles cannot be adequately understood without also trying to grasp the symbolic dimensions associated with them. In this sense, it is too limited to understand railways and automobiles only in terms of their technological benefits as a means for overcoming spatial distances. While it is undoubtedly true that the car had advantages over the tram and regional railways in the 1920s, as the car was much more flexible in terms of time and space than the railways, which were linked to timetables and rail, rational reasons alone do not make plausible the *explosive* and highly *sustained turn* to the car by the Americans, and the Angelenos in particular. For example, they do not explain why the negative consequences of this technology, in the form of accident fatalities, air pollution, landscape consumption, and congestion, were accepted as more or less inevitable. Transport modes, as will be shown, have symbolic dimensions[17] which influence both individual behavior and transport policy. In the context of this work, the symbolic dimensions of both the car and the railways are outlined. The final question is whether the reorientation of transport policy from freeway to railway construction led to a change in the understanding of mobility.

---

[17]Like signs in general, symbols stand for something. However, as signs of a special kind, they also have some special features:

- Symbols have an excess of meaning in the sense that they change original sign meanings.
- Symbols condense complex relationships with particularly simple, easily comprehensible signs (cf., e.g., Edelman 1990: 27 ff).
- Symbols trigger not only cognitive but also affective signals in their users. They thus provoke normative evaluations and perspectives for action.
- Finally, symbols are particularly context dependent. Their meaning can vary considerably, depending on the membership of individuals and groups, the context of use, language, culture, and especially time (cf. Prittwitz/Bratzel/Wegrich 1992: 1 ff; the concept of the sign is also discussed in Eco (1977: 25 ff and 190 ff); the concept of the symbol is also dealt with by Pross (1974: 23 ff), Meyer (1992: 50–55) and Voigt (1989), among others).

6.3.3.1 *Cars and freeways as symbols*

The public fascination with the automobile in the 1920s and long after the Second World War cannot be explained by utilitarian factors alone. Rather, the automobile — far more than the "uncomfortable and crowded trains" — promised to maintain or even expand the American core value of individualism, defined as privatism, and "freedom of choice and opportunity", in a changing urban–industrial society. The automobile increased the spatial mobility potential of car owners and thus also strengthened the American core value of individual freedom. Thus, the term *freeway* itself is already associated with individual *freedom*. The historian John D. Hicks also pointed to the connections between sociocultural factors and cars, noting for the 1920s that "this automobile psychology seemed to characterize the nation as a whole; the American people, like the drivers of many American cars, were relentlessly on their way, but not quite sure where they were going, or why" (quoted after: Brownell 1972: 21).

The social position in American society is very strongly linked to the place of residence (see above). The wide spatial separation of workplace and suburban residence is also an expression of the (preferred) lifestyle of the citizens of Los Angeles. Spatial mobility is not only a basic prerequisite for realizing this lifestyle but also a demonstration of social status. In this sense, the journey from the workplace in the city to the suburban residence is, beyond mere commuting, "a drive for achievement":

> The deeply ingrained cultural drive for achievement nurtures a vigorous pursuit of higher socioeconomical status. It is expressed geographically by spending one's earnings on the most expensive home affordable in the best possible residential area. Once one attains a higher-status mode of living, one joins in efforts to protect the neighborhood against the entry of status-challenging lower income groups. Should these efforts to defend against the down-grading of the neighborhood fail, one may find it necessary to pull out in order to avoid the social stigma of eventually being identified with lower-status newcomer (Muller 1981: 65).

The car is therefore inevitably valued highly by society, as its ownership enables citizens to live in the better residential areas of the suburbs. This is also the opinion of Flink: "In our traditionally mobile society the motorcar was an ideal status symbol" (1975: 38). In their "Middletown" studies of the 1930s, Robert S. and Helen Lynd also emphasized the important social significance of the new mode of transport, as the working class in particular "want what Middletowns wants, so long as it gives them their great symbol of advancement — an automobile. Car ownership stands to them for a large share of the *American dream*; they cling to it as they cling to self respect" (1937: 265 f). In the 1920s, the automobile rose to become a "symbol of modernity" (Brownell 1972), in which everyone wanted to participate.

The early and extreme turning to the car in Los Angeles can be deduced not least from the — at that time — unique distance-intensive mobility structure. Due to the low settlement density and the rapid decline of the railways, car ownership and social status were inseparably linked. Those who did not own a car could hardly participate in the suburban lifestyle. In the car metropolis of Los Angeles, social status was defined even more vehemently by the automobile than elsewhere:

> A car was the absolutely essential piece of social overhead capital (...)
> To have a car meant being somebody; to have to borrow a car meant knowing somebody; to have no car at all, owned or borrowed, was to be left out — way out (Wilson 1967: 40).

This coupling of automobility and sociocultural factors makes it easier to understand why citizens were so irrationally attached to it. In the 1920s and even decades later, traffic deaths and traffic jams were not seen as being directly caused by the positive symbol of the automobile. Rather, the problem was mainly seen in technical shortcomings, such as inadequate traffic regulations and compliance with them, and a lack of road capacity (cf. above; Brownell 1972: 28 ff). Hardly any other way than through the affective surplus of meaning of symbols is it understandable why these victims in the form of deaths, among other things, were (and are) accepted or why no measures with a high depth/sharpness of effect were (and are) taken. As an explanation, it is often pointed out in this context that automobility has developed religious forms (e.g., Sloterdijk 1992). In Lewis Mumford's view, too, the symbol of freedom, the automobile, has practically developed into a religion "and the sacrifices that people are prepared to make for this religion stand outside the realm of rational criticism" (1964: 244 f). David Brodsly gave a similar assessment when he described the freeways as the modern cathedrals of L.A.:

Every time we merge with traffic we join our community in a wordless creed: belief in individual freedom, in a technological liberation from place and circumstance, in a democracy of personal mobility. When we are stuck in rush-hour traffic the freeway's greatest frustration is that it belies its promise (1981: 5).

Brodsly thus addressed the dissonances that increased in the 1970s, pointing to a symbolic change of cars and freeways. The meaning of the car symbol had already changed once. At the beginning of the 20th century, the automobile — as a "toy of the rich" — was still quite negatively associated. In 1906, the future American President Wilson stated that the possession of a motor car "was an obtrusive display of wealth, that the envy of pedestrians must necessarily lead to socialist resentment" (quoted from Heimann 1987: 19).

Only a few years later could a change of symbols be observed. The car was given positive connotations, and the solution to the complex problems of the metropolis was condensed into the symbol of the automobile. Experts and "the man in the street" equated the promotion of this mode of transport with the solution to the city's problems. As early as the end of the 1920s, car manufacturer Henry Ford had already pointed out the simple way to solve urban problems: "We shall solve the city problem by leaving the city" (quoted after: Flink 1975: 39).[18] Since the 1920s, the car seemed to people to be a means of driving away from the problems. And this — paradoxically — despite the extremely worsening congestion problems caused by increasing car traffic.

These positive associations were extremely stable among broad sections of the population at least until the 1960s. Since the 1970s, and increasingly since the 1980s, the car as a positive symbol had again been undergoing change. It became increasingly clear that the Freeways had not kept the promises of their supporters. The L.A. Freeways were increasingly associated with "all the ills of a modern, and particularly automotive, metropolis: air and noise pollution, congestion, the destruction of neighbourhoods, the specter of a concrete blanket over the landscape" (Brodsly 1981: 51 ff; cf. also Wachs 1992). Because of these problems, Los Angeles became a negative symbol of an automotive

---

[18]Following this pattern, "(T)he ultimate answer to the tenement house slum (...) (was) that everyone should buy a motorcar and commute to suburbia, and a projected suburban real estate boom soon became another anticipated benefit of automobility" (Flink 1975: 39).

society and a failed transport policy, especially in other countries (cf., e.g., Monheim/Monheim-Dandorfer 1990).

In Los Angeles, too, it became clear that "(F)ew will pay for new roads" (Seelye 1984). In the early 1980s, for example, six Californian counties rejected tax increases for road projects in referenda by an overwhelming majority.[19] In contrast, tax increases for public transport in Los Angeles and other counties were supported by voters (see *ibid.*). As outlined above, the LACTC began building a gigantic rapid transit system in the 1980s, on which many billions of dollars were to be spent. In 1990, the light rail line from the CBD to Long Beach (Blue Line) was opened; in 1992, sections of the commuter rail system (Metrolink) were put into operation; and in early 1993, underground trains also started running (Red Line).

6.3.3.2 *Light railways as mobility symbols*

In Los Angeles, the railways are once again associated with solving the mobility problems of the metropolis. As early as the middle of the 19th century, the railway was the symbol of progress par excellence. The advent of the railway was accompanied not only by a considerable increase in spatial mobility but also by economic growth. Corruption, poor service, and little comfort led to the railways being viewed very negatively by the population, at least since the 1920s. At that point, the car symbolized progress and mobility and was expected to solve the problems of the city. However, soon after the discontinuation of operations at the beginning of the 1960s, the well-known phenomenon of a "mythical" transfiguration of the railways set in with increasing time lag (Richmond 1991). In the early 1970s, Snell drew the (apparently false) picture of a *light rail utopia* in Los Angeles, which General Motors destroyed in a conspiratorial act.[20] The railways were once again positively reputed.

---

[19] In an interview, Richard Barret (Principal Planner), of the Automobile Club of Southern California, also explained that the previously overwhelming popular support for the construction of highways and freeways had now declined sharply (interview on 10 November 1992, Los Angeles).

[20] "Thirty-five years ago Los Angeles was a beautiful city of lush palm trees, fragrant orange groves and ocean-clean air. It was served then by the world's largest electric rail network. In the late 1930s General Motors and allied highway interests acquired the local transit companies, scrapped their electric trains, tore down their power transmission lines,

In the face of daily smog and traffic jams on the streets, there was increasingly a sounding board that put the railways in a new light. As was previously the case with cars and freeways, hopes of solving urban problems were now growing in the railways:

> Whether one's concern was the economic vitality of cities, protecting the environment, stopping highways, energy conservation, assisting the elderly and handicapped and poor, or simply getting other people off the road so as to be able to drive faster, transit was a policy that could be embraced. This is not to say that transit was an effective way of serving all these objectives, *simply that it was widely believed to be so* (highlighted, S.B.) (Altshuler 1979: 36).

In Los Angeles, the symbolic elevation of the railways could be an explanation why, contrary to scientific analysis, the metropolis is investing exorbitant sums in a railway network. The preliminary assessment of the railway lines that have been in operation since 1990, as attempted in Section 4.2.1, showed that the metropolis could have been much better served by an extensive bus network, both in terms of efficiency and fairness. The most important reason for this was the mobility structure of the metropolis, which is characterized by a low settlement density and a dispersion of jobs. The resulting dispersal of destination and source traffic makes a high-capacity mode of transport, such as rail, inefficient. In Section 5.2.2, an additional attempt was made to point out that the dominant supply-side orientation of the metropolitan transport policy has only a limited depth of impact. Nevertheless, the solution to many of the metropolis's problems was associated with this mode of transport.

As Jonathan Richmond vividly demonstrated in his dissertation "Transport of Delight — The Mythical Conception of Rail Transit in Los Angeles" (1991), the symbol of the railway condenses hopes for the solution of traffic and other urban problems. Compared to buses, railways not only appear to be more comfortable, safer, more environmentally friendly, and more suitable for solving the metropolis's congestion problems but the construction of railways is also "(t)he Symbolic Promise of Community Renewal" (Richmond 1991: 271–290).

---

ripped up their tracks, and placed GM buses on already congested Los Angeles streets" (1974: 3).

It has already been discussed that the problems of economically dis-advantaged areas of the city and its inhabitants are not primarily transport related. Poor education and training, violence, racism, and unemployment are the root causes of their economic situation. The light rail system also brings them little benefit because it "does not serve the principal employ-ment destinations of the community, nor the needs of local trips in gen-eral" (*ibid.*: 289). And further, "There is no evidence, furthermore, that the arrival of rail will stimulate development, given the lack of intrinsic attraction to developers to invest in these communities" (*ibid.*: 293).

One possible explanation why the railways are nevertheless perceived as a solution to the problems lies in their positive symbolic value as "transport of delight: a symbol of progress at which all can marvel, what-ever the reality of its actual performance in enhancing mobility, alleviat-ing congestion, or reducing pollution" (Richmond 1991: 291). The symbol has met with the necessary response from the public precisely because the railway — similar to the car before it — can be linked to posi-tive associations which lie beyond the objective (problem-solving) dimen-sions. Because, as Richmond pointed out,

> The rail system also symbolizes technological virtuosity and sex appeal; it carries nostalgic, romantic and religious connotations and is a symbol of community pride, independently of any transportation advan-tages it may provide (...) It derives that meaning from a rich context of historical associations, and from the experiences and memories of those for whom it seems desirable (1991: 4; 263).

After several defeats in referenda, plans were made to extend the railway network extremely across the metropolis. This also seemed to accommodate the "low-density" lifestyle of the population and thus did not trigger *cognitive dissonances* (Festinger) with other sociocultural orientations.

The symbol concept makes the irrational adherence of transport pol-icy to rail-bound public transport in Los Angeles somewhat more under-standable. People try to reduce complexity because it is cognitively unpleasant and can cause fear and insecurity, among other things. They therefore resort to simplifications such as symbols.[21] In the case of L.A.'s

---

[21] As Edelman put it, "It is characteristic of a large number of people in our society that they think and see in stereotypes, personalisations and simplifications; that they cannot perceive or bear ambiguous and complex situations; and that they react accordingly mainly

transport policy, the symbol "railway" reduces complexity by offering a simple technological answer to complex social problems, thus fulfilling the human need for simplicity. If it is true that the population needs symbols to get excited about policies, then L.A.'s fixation on motorized modes of transport becomes more explainable. Cars, freeways, and even railways have a much *more tangible* "symbolic value" than, for example, restrictive land-use regulations, traffic taxes, or transport demand management measures, even if the latter have a much greater depth of effect.

6.3.3.3 *Is there a change of values and attitudes in Los Angeles?*

Although one can certainly speak of a symbolic change, the surveys available to me and the above analysis of mobility patterns do not indicate a significant change in the understanding of mobility. Values and attitudes regarding mobility remained unchanged, with hopes being transferred only on a *new* mode of transport. For the most part, citizens are not prepared to change their mobility behavior or lifestyle. In detail, the picture is as follows:

It is generally known that urban sprawl increases the demand for cars and is considerably more energy and traffic intensive than dense structures (Cervero 1991: 479; Altshuler 1979: 379 ff). Despite this consensus, the population's preference for a suburban lifestyle has remained unbroken for decades (cf. above). This preference is expressed both in opinion polls (cf. Altshuler 1979: 377 f) and in actual behavior (Müller 1981). The suburban population even tries to resist developments which could lead to a densification of the suburbs and thus also to a reduction in distance-intensive mobility behavior (Baldassare 1986). The tendency toward job dispersion has indeed led to a "commuting paradox" in some metropolitan areas (Gordon *et al.* 1991), in so far as average commuting times have shortened somewhat between 1980 and 1985. Nevertheless, the spatial living and working imbalance has continued, so commutes continued to increase (Cervero 1989b). Distances for "non-work-trips", which are already booming, have also increased in recent years (Gordon/Richardson 1992). In this respect, no reorientation of suburban lifestyles and mobility behavior can be perceived.

---

to simplifying symbols" (1990: 29). Festinger (1978) also pointed out in his "Theory of Cognitive Dissonance" that the human tries to avoid complexity and cognitive contradictions.

A paradoxical picture emerges regarding transport policy attitudes. Although the citizens of the suburban Orange County have considered the traffic situation to be one of the most important problems since the 1980s at the latest, and since 1985 the most important problem of the metropolis (Baldassare 1991: 212), this insight has not led to changes in behavior (cf. above; Baldassare 1991: 214). Rather, Baldassare stated,

> There is substantial opposition to transportation policies that involve financial or lifestyle sacrifices, despite the experience of worsening commutes and the growing perception of traffic problems. This is evident in survey responses to questions involving local tax increases and the recent proposals to reduce automobile driving (1991: 216).

After all, in 1980 and 1990, the people of L.A. County voted in favor of a VAT increase, much of which benefited a rapid transit system. But how can this approval be explained by the resistance to change in behavior? As early as 1974, a Los Angeles Times survey made the surprising discovery that while just under 87% of those questioned thought L.A. needed a new rapid transit system, only just under 5% replied that they would actually use it (Marcuse 1975: 1). This has not changed, despite a majority vote in favor of the construction of a vast network of express trains. When car-driving commuters were hypothetically asked whether they would take the train to work "just to try it out and see if they liked it", 71% replied in 1991 and 85% in 1992 that they "*would not even try it*". Just 11 and 6%, respectively, were sure that they would test the alternative mode of transport at least once (Commuter Transportation Services 1992: 46 f). This "paradox" (approval of the construction of rapid railways — rejection of the use of rail) could be explained by the fact that citizens hope that others will make use of the new mode of transport and relieve the burden on the freeways so that they themselves can travel faster again.

At least some exogenous events can change the strong mobility attitudes and travel patterns if only in the short term. It was not until the Californian earthquake of January 1994, for example, that many commuters in Los Angeles were forced to test alternatives to driving their cars to work. For the first time, many Angelenos changed their mobility behavior and filled the commuter trains of the Metrolink, although many switched back to the car after the repair of the freeways.

Chapter 7

Back to the Future: L.A. Revisited 1990–2022

We cannot solve our problems with the same level of thinking that created them.

Albert Einstein

Has traffic in Los Angeles improved over the past 30 years? To what extent have the policymakers in the region learned lessons from the strongly supply- and capacity-oriented transport policy programs of the past and developed new strategies for transport and mobility? And how will the new technological trends of autonomous driving affect the traffic situation and mobility structures? These questions are the focus of this chapter. Against the background of the theses of the first edition of the study in 1995, the developments in transport policy in Los Angeles over the past decades were examined.

7.1 Traffic and transport trends in Los Angeles in the past three decades

Los Angeles is still one of the most congested cities in the world. According to the Inrix Global Congestion Ranking, "Los Angeles topped the list of the world's most gridlocked cities for the sixth straight year, with drivers spending 102 hours in congestion in 2017 during peak time periods" (Inrix 2017). According to the TomTom Traffic Index, the time

131

lost due to traffic jams during rush hour in 2019 (i.e., before the COVID-19 pandemic) was as much as 167 hours, i.e., 6 days and 23 hours. A 30-minute journey was then extended by 19 and 25 minutes, respectively, at peak times in the morning and evening.[1] The lockdown only seems to have brought temporary relief. After the end of the lockdown in June 2020, the traffic volume rose again to the pre-pandemic level.[2]

A major driver of the deteriorating traffic situation in the region is the increasing population. Between 1990 and 2020, the population of the greater Los Angeles area grew from around 14.5 to around 19 million people. Los Angeles is one of the fastest growing regions in the US. At the same time, the volume of traffic has continued to rise. More than 16 million vehicles are registered in the metropolis with around 12.9 million driver license holders. With 844 cars per 1,000 inhabitants, the vehicle density is one of the highest in the world (see Table 7.1).

Table 7.1: Population and automobile density in metropolitan Los Angeles (2020)*.

Population	19.2 million
Registered Vehicles	16.2 million
Licensed Drivers	12.9 million
Car Density (vehicles/1,000 residents)	844

Notes: *Los Angeles–Long Beach–Riverside Combined Statistical Area. Southern California Association of Governments (SCAG) is a Joint Powers Authority under California state law. The SCAG region encompasses six counties (Imperial, Los Angeles, Orange, Riverside, San Bernardino, and Ventura) and 191 cities in an area covering more than 38,000 sq. miles.
Source: SCAG (2021, 2022).

---

[1] Traffic Index of TomTom is defined as "extra travel time during peak hours as compared to a one-hour period during free flow conditions, multiplied by 230 working days per year" (cf. TomTom 2022).

[2] "On June 30, daily traffic in Los Angeles and Orange counties was above pre-pandemic levels by 4%, according to the traffic data company Inrix. Traffic during peak morning hours was 86% of pre-pandemic levels for Wednesdays, while traffic during peak afternoon hours was the same as pre-pandemic levels." (Schnalzer 2021). The most extreme increase in traffic based on the June 30 data was observed during off-peak hours, during which Inrix found traffic to be 14% higher than pre-pandemic levels. This may point to the flexibility of people who are still working remotely (cf. *ibid.*).

The motorization trend is also unbroken in Los Angeles County, the most densely populated core area of the metropolis. Between 1990 and 2020, the number of vehicles (cars/trucks) increased from 5.9 to 7.5 million (+27%), while the population grew by only 13% (cf. L.A. Almanac 2022). The motorization rate or car density increased in this period from 663 to 747 vehicles per 1,000 inhabitants.[3]

In 2020, the privately owned cars remained the dominant means of transport in the metropolis of Los Angeles. Around 86% use the private car to get to work, with 76% driving alone and 10% carpooling. Only 4% of commuters use public transport to get to work, and another 4% walk or cycle. 6% of employees work from home (cf. SCAG 2021).

In the more densely populated Los Angeles County, the figures differ only slightly: In 2020, almost 81.6% drive to work (72% of them are single drivers), while only 5.4% use public transport and 8% work from home. Hardly anything has changed in the last 30 years: In 1990, 85.6% were car commuters (70.1% of whom were single drivers) and 6.5% were public transport commuters, while only 2.7% worked from home. Home or telework alone has changed significantly between 1990 and 2020 (see Figure 7.1).

7.2 Rail infrastructure expansion: A new era of transport and mobility policy?

Basically, since the beginning of the 1990s, the supply-oriented mobility and transport policy in Los Angeles has been continued in various respects. But since the 1980s, there has been an increasing trend toward the expansion of rail transport infrastructure.

However, in the last 30 years, the size and capacity of the freeway and street network have also continued to increase, albeit to a lesser extent than before. Central road infrastructure projects concern, for example, the 405 freeway, one of the busiest and most congested freeways in the USA. In 2014, a one-lane extension of the 405 Freeway's Sepulveda Pass opened after five years of construction. As with many

---

[3] In addition, metropolitan L.A. continues to be one of the regions with the worst air quality in the country: "According to the 2019 State of the Air report, which compared data across 229 metropolitan areas, Los Angeles has the worst ozone air pollution in the United States" (IQAIR 2022).

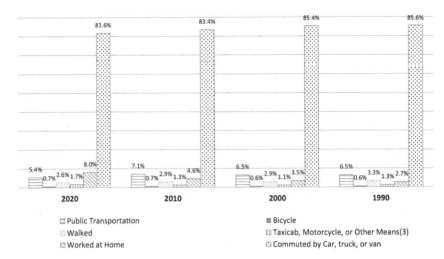

Figure 7.1: Commuting by workers — Los Angeles county (workers aged 16+ years).

Source: Data from US Census Bureau.

other capacity expansions, a study by the Los Angeles County Metropolitan Transportation Authority found that just one year after the opening, travel times during rush hour increased by another minute on average (McCarty Carino 2018).

Regardless, the capacity of the 405 Freeway continues to increase elsewhere: In 2018, construction began on a freeway extension in Orange County which will be a 16-mile stretch of the highway from State Route 73 in Costa Mesa to Interstate 605 where one lane will be added in both directions and two existing carpool lanes will be converted to a toll system. At $1.9 billion, it is claimed to be the second most expensive highway project in state history. And again, the responsible authorities argue that by expanding the freeway, significant time savings could be achieved by reducing traffic jams.[4]

However, the supply and capacity expansion of roads and freeways in Los Angeles are encountering increasing resistance. Some large-scale car-centric projects are also being abandoned due to demands for

---

[4]Caltrans Director for Orange County, Ryan Chamberlain, comments: "Currently the 405 is the most heavily congested freeway in the nation, 370,000 daily vehicle trips on that corridor. And in 2040 or close to 2040, with this improvement, we're anticipating an hour and a half — or more — in travel time savings for most commuters" (KPCC 2018).

climate-friendly alternatives. In May 2022, for example, the extension of the 710 corridor in Los Angeles, a key route from the ports of Los Angeles and Long Beach to distribution centers, was officially canceled by the Los Angeles Metropolitan Transport Authority (Metro). The original plan, which dates back 20 years, was to expand the freeway from 8 to 16 lanes (Brey 2022; cf. Ionescu 2022).

In contrast, over the past 30 years, rail transportation in Los Angeles has experienced a renaissance while maintaining the supply-side mobility pattern. After the abovementioned reopening of the first light rail connection in 1990 (Blue Line), the rail transport capacity in the metropolitan area of Los Angeles expanded significantly. The Blue Line (later named "A Line") was followed by the Red Line (1993–2000), Green Line (1995), Gold Line (2003), and the Expo Line (2012/2016) as well as the Crenshaw Line (K Line, 2022). In addition, there are two express bus lines, the Orange and Silver Lines, on special lanes (busways), which the Los Angeles Metropolitan Transport Authority (Metro) describes as "light rail on rubber tires". Overall, the public rail transport network grew between 1990 and 2022 to a total of 113.5 miles (182.7 km) with 99 stations (see Table 7.2).

Table 7.2: New construction and expansion of metro rail and busways (1990–2022).

Rail Line	Opened	Miles	Type	Stations	Construction cost
Metro Blue Line (A Line)	1990	21.3	Light Rail	22 (inc. 3 shared)	$877 million
Metro Red (B Line)/ Purple Lines (D Line)	1993 MacArthur Park, 1993 Wilshire/Western, 1996 Hollywood, 1999 North Hollywood, 2000	14.0	Subway	16 (inc. 6 shared)	$4.5 billion
Metro Green Line (C Line)	1995	19.5	Light Rail	14 (inc. 1 shared)	$718 million
Metro Gold Line (L Line)	2003 Eastside Extension, 2009 Azusa Extension, 2016	29.7	Light Rail	27 (inc. 1 shared)	$2.8 billion
Metro Orange Line (G Line)	2005 Extension from Canoga Park to Chatsworth, 2012	18	Busway	18 n/a	$484 million

(Continued)

Table 7.2:   (*Continued*)

Rail Line	Opened	Miles	Type	Stations	Construction cost
Metro Silver Line (J Line)	2009 South Bay and El Monte via Downtown Los Angeles	n.a.	Busway	11 n/a	$587 million
Metro Expo Line (E Line)	2012 Extension to Santa Monica, 2016	18.1	Light Rail	19 (inc. 2 shared)	$2.4 billion
Crenshaw Line (K Line)	2022 Expo/Crenshaw — Downtown Inglewood — Westchester/ Veterans	8.5	Light Rail	7	$2 billion

Source: Data from Metro (2022).

An important enabler for the construction of the new railway network was and is the VAT increases ("Measure R", "Measure M") approved by the voters in 2008 and 2016, with which the (partial) financing was made possible. The policymakers in Los Angeles highlight that "in 2016, Angelenos showed their overwhelming support for the transformation of our public transportation system with the approval of a no-sunset sales tax, Measure M, that will provide $120 billion over 40 years to improve the way Angelenos travel in the region" (City of Los Angeles 2022: 26).

Many major rail projects in L.A. County are currently under construction or in the pipeline, such as the Metro Purple Line Extension and the Foothill Gold Line Extension. In January 2022, the "West Santa Ana Branch Transit Corridor Project" was decided, but according to current planning, its opening would not be until 2043 (Uranga 2022). Planning is already well underway for many other rail construction projects, such as the East San Fernando Valley Light Rail Transit Project or the Eastside Transit Corridor Phase 2. The planning horizon for the rail network expansion through the metro in Los Angeles now extends to the year 2067 (see Table 7.3).[5]

---

[5]Due to the rapidly increasing costs, many local rail transport projects have already been divided into phases in order to start implementation. "Plans for transit in the San Gabriel Valley and other communities in the eastern side of L.A. County have undergone numerous changes since the passing of Measure M in 2016. Plans for a two-pronged L Line extension were pruned in 2020, when Metro dropped a proposal to build a second alignment parallel to the SR-60 Freeway, citing both the high cost and the operational

Table 7.3:   Projects under construction/planned projects in Los Angeles.

Rail Line	Construction start	Miles	Type	Opening (planned)	Construction cost
Metro Regional Connector Project	2014	1.9	Subway link connecting A (Blue), B (Red), and L (Gold) Line	2023	$1.76 billion
Metro Purple (D) Line Extension	2015	9.1	Subway	2024/2025/ 2027	$8.79 billion
Foothill Gold Line Extension	2020	9.1	Light Rail	2025	$1.5 billion
Metro Airport Connector Project	2021	—	Multilevel rail station	2024	$0.9 billion
West Santa Ana Branch Transit Corridor Project	2023	19.3 (Alternative 1)	Light Rail	2043	$8.5–8.8 billion
East San Fernando Valley Light Rail Transit Project	2023	9.2	Light Rail	2028–2030	$2.8–3.6 billion
Eastside Transit Corridor Phase 2	2025	9.0	Light Rail	2035	—
Extension of Line C to Torrent	—	—	Light Rail	2030–2033	—
Sepulveda Pass Transit Corridor	—	—	Light Rail/Subway	2033–2035	—
Crenshaw Northern Expansion	—	—	Light Rail	2047	—
Vermont Transit Corridor	—	12.5	Subway	2067	—

Sources: Fine (2021), Sharp (2022), and Metro (2021, 2023).

Overall, with the planned projects, Los Angeles is currently the region with the largest rail expansion program in the US. Given the outlined mobility history of L.A., which in the 1920s already operated the largest regional rail network in the world and gave it up completely by the early 1960s in favor of the construction of a huge freeway network, the enormous boom in local rail transport is another ironic twist in transport policy of the metropolis (Finc 2021).

How has the improvement in supply affected transport demand in Los Angeles? The interim balance after more than 30 years of light rail and subway construction is comparatively sobering. As mentioned, despite the significant investments in public rail transport, car density in Los Angeles has continued to increase sharply. And travel behavior has not changed significantly in favor of public transport.

However, due to the expansion of the local rail network, the passenger volume initially increased moderately from around 400 million trips at the beginning of the 1990s to around 480 million trips in 2010. After a phase of stagnation between 2010 and 2013, passenger numbers fell sharply again. In the five-year period between 2014 and 2019 — before the COVID-19 pandemic — the figures showed a slump in passenger numbers from 476 to 376 million, which corresponds to a decline of more than 20% (see Figure 7.2).[6]

The sharp decline in passenger demand triggered a kind of "transit blues" that surprised decision-makers and triggered controversial discussions in public (Wassermann/Taylor 2022) because, with the consistent expansion of local rail transport and the simultaneous high population growth in L.A. County, there was in principle even greater passenger potential available for public transport. What are the possible reasons for the drastic decline in passenger numbers?

7.3 Reasons for the "transit blues" in Los Angeles

Studies indicate that various influencing factors played a role in the development of the transit blues. A crucial explanatory factor, however, is the

---

complexities that would be created. (...) This is not the first Metro Rail expansion to recently be pared back as a consequence of ballooning costs" (Sharp 2022).

[6]Wasserman and Taylor (2022) note a similar development for Greater Los Angeles: "Between 2014 and 2018, Greater Los Angeles lost 128 million annual riders, a drop of 18%" (Wasserman/Taylor 2022: 4); see also Miranda (2020: 13).

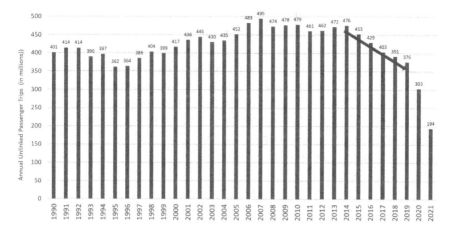

Figure 7.2:   Los Angeles county metropolitan transportation authority: Annual ridership trends.

Sources: Data sources: Rubin/Moore (2019: 4), and Statista (2022).

increase in vehicle ownership. Between 1990 and 2000, there were four new residents for every additional vehicle (0.25 vehicles per new resident). In contrast, between 2008 and 2018, there was one newly registered vehicle for every new resident.

Vehicle ownership increased across all income brackets. But growth in passenger cars was particularly strong among low-income households, i.e., among the very groups in the population that had so far been the main user group for public transport. 78% of public transport (Metro) passengers do not have access to a vehicle — mostly for economic reasons — and are largely dependent on public transport (Manville *et al.* 2023: 308, 2018: 68). In fact, between 2008 and 2018, the proportion of households without a car in the Los Angeles Metropolitan area fell from 7.6% to 6.6% (Rubin/Moore 2019: 3). During the same period, the economic environment in the region improved, which can be seen in the growth in employment and real household income. At the same time, car costs have fallen significantly: Above all, the fuel costs, which had previously risen sharply, have fallen steeply again since 2012 (Manville *et al.* 2023: 321).

Due to the higher number of vehicles owned by households, many previous users of public transport — like the vast majority of Angelenos — were increasingly able to switch to the car for their mobility needs. The private automobile is much more suitable for the long-distance mobility structures of Los Angeles. Despite increasing traffic congestion, the accessibility by automobiles compared to transit is much higher in Los Angeles. Studies show that the average number of jobs that could be reached in 30 minutes was 55 times higher by automobile than by transit in 2014 (*ibid.*: 309)![7] In addition, there are usually enough free parking spaces available in Los Angeles at the point of departure as well as at the destination.[8] "Driving is relatively easy, while moving around by means other than driving is not. These circumstances give people strong economic and social incentives to acquire cars, and — once they have cars — to drive more and ride transit less" (Manville *et al.* 2018: 10).

The distance-intensive spatial structures and the associated car-oriented mobility culture mean that public transport is not perceived seriously or at all as a real mobility alternative. Despite investments in the rail infrastructure and an increased network, the metropolis of Los Angeles has not succeeded in increasing the attractiveness of public transport for broad sections of the population.[9] Analyzes show that only population groups with very low incomes and those who do not own a vehicle use the public transport service far above average. It is somewhat ironic that as their income increased, they immediately turned their backs on buses and trains, and as a result the number of passengers dropped drastically in the 2010s.

One lesson from this is that it makes no political sense to envisage public transport primarily as a social project and a catchment area for residents who have no alternative mobility option. According to surveys, only 2% use the buses and trains in Los Angeles very frequently, and around 20% occasionally. In contrast, 77% of the approximately 19 million

---

[7] Interestingly, job access by automobiles had fallen in 2017 mainly due to congestion, but the auto–transit ratio is still 33:1 (cf. Manville *et al.* 2023: 309).

[8] Sorensen *et al.* (2008: 78) point out, "(...) the supply of parking in the region — as a result of the long-standing zoning policy of minimum parking requirements — is much higher than one would expect in a region with the density of Los Angeles. This reduces the average price for parking, which, in turn, encourages more travellers to drive."

[9] "The new service added was not attracting ridership. Indeed, service-effectiveness (measured in terms of boardings per service hour) fell 22% between 2014 and 2018 in Greater Los Angeles. (...) Amidst the ridership downturn, operators in Greater Los Angeles spent $5.3 billion in 2018, around 1.5 times more than a decade earlier. A substantial share of this investment went into new rail transit service in LA County" (Wasserman/Taylor 2022: 8).

inhabitants of the region rarely or never use the trains and buses (Manville *et al.* 2023: 307–308, 2018: 68).

A good example is the construction of the subway (Red Line) with a total cost of 4.5 billion dollars for a route of 14 miles and 16 stations. 20 years after its opening, the following comment was made in a publication by the Los Angeles Metropolitan Transport Authority: "One of the peculiarities of the subway is that even today many residents in parts of the county have never used it or do not know it exists" (Sotero 2013). The attractiveness of public transport also depends on qualitative factors such as a sense of security or discomfort, for example, because of homeless people on trains or stations. In fact, passenger surveys of Metro in 2018 indicate that a sense of security was also a reason for some to drive less or not at all (Manville *et al.* 2023: 320).

There are also other reasons for the declining passenger numbers. This includes the emergence of ride hailing by the Transportation Network Companies (TNCs), especially Uber and Lyft, which have established themselves in Los Angeles as a cheap and reliable mobility alternative since 2013. In contrast to public transport, they pick up the passengers at the starting point and drive them directly to their destination, eliminating the need for transfers and footpaths. For many, this represents both a gain in comfort and safety and, in view of the spatial structures, is a clear advantage over public transport.[10] Studies show, however, that in Los Angeles — in contrast to the decline in public transport passengers in the San Francisco region — they are of very little importance (cf. Wasserman/ Taylor 2022: 9–10).

7.4 New sustainability and transport goals in the city of Los Angeles

The transport sector has recently received further dynamism from the sustainability plans of the City of Los Angeles. Based on the commitment to the Paris Agreement, the "New Green Deal Pathway" presented in 2015 and further detailed in 2019 is a sustainability plan with five zeros as its goal: zero-carbon grid, zero-carbon buildings, zero-carbon transportation,

---

[10] Studies show that, on the one hand, they also generate new traffic through journeys that were not previously made. On the other hand, they also act as direct competitors to other modes of transport, especially pedestrian and public transport (cf. Graehler *et al.* 2019).

Table 7.4:   Sustainability plan 2019: Targets mobility & transportation.

- Increase the percentage of all trips made by walking, biking, micro-mobility/
  matched rides or transit to at least 35% by 2025; 50% by 2035; and maintain at
  least 50% by 2050
  Baseline: 14% of all trips made by non-car modes in 2015
  Source: 2016 City of Los Angeles Travel Demand Forecasting Mode
- Reduce Vehicle Miles Traveled (VMT) per capita by at least 13% by 2025; 39% by
  2035; and 45% by 2050
  Baseline: 15 VMT per capita per day
  Source: Southern California Association of Governments Transportation Demand
  Model
- Increase the percentage of zero emission vehicles in the city to 25% by 2025; 80%
  by 2035; and 100% by 2050
  Baseline: 1.4% of vehicles as of September 2018
  Source: CA Department of Motor Vehicles
- Electrify 100% of Metro and LADOT buses by 2030
  Baseline: 4.6% L.A. Metro (zero in service), 2018; 8.2% LADOT (four in service),
  2018. Includes buses on order
  Source: L.A. Metro, Los Angeles Department of Transportation
- Ensure Los Angeles is prepared for Autonomous Vehicles (AV) by the 2028
  Olympic and Paralympic Games
  Source: City of Los Angeles (2019)

zero waste, and zero wasted water. These targets are backed by more than 400 initiatives aiming to lead Los Angeles to carbon neutrality by 2050 (City of Los Angeles 2019: 70, *ibid.* 2022).[11]

High targets have been set for the transport sector for deep reductions in greenhouse gas emissions by 2050 (Table 7.4). Mode shift should be achieved by improving the transportation system to enable Angelenos to use public transit and other modes to get where they need to go. "This is key to going carbon neutral because mode shift will allow us to reduce today's transportation emissions by a quarter, equivalent to removing 300,000 cars from the road for one year" (City of Los Angeles 2019: 70).

Among other things, the plan stipulates that the proportion of journeys made by public transport, bicycle, micromobility, or on foot should

---

[11]"Based on our commitment to the Paris Agreement, this plan charts a new course for Los Angeles' emission reduction targets — the 2019 Green New Deal Pathway — which calls for cutting greenhouse gas emissions (GHGs) to 50% below 1990 levels by 2025; 73% below 1990 levels by 2035; and becoming carbon neutral by 2050." (City of Los Angeles 2022).

increase to at least 35% in 2025 compared to 14% with non-car modes in 2015. At the same time, the vehicle miles traveled (VMT) per capita should be reduced by at least 13% by 2025 and 39% by 2035. The average distance by car is 15 miles per capita per day.

A number of programs and measures with milestones are planned to achieve the goals in the "Mobility & Transportation" area. The following important milestones have been set:

- Complete Measure M 28 (e.g., subway and light rail network expansion) (2028).
- Improve travel time on L.A. County's bus network by 30% (2028).
- Support Metro with the implementation of a congestion pricing pilot (2025).
- Ensure all city residents have access to high-quality mobility options within a 10-minute walk from home (e.g., bike lanes, electric car sharing) (2028).

Accordingly, the 2021/2022 Annual Report of the Green Deal points out that Los Angeles has implemented the nation's largest-in-history public infrastructure program, Measure M, to fund transit projects in the region indefinitely. It also mentions that the city launched the NextGen Bus Plan to reimagine the region's bus system to create faster, more frequent, reliable, and accessible service, and that Los Angeles installed over 25 lane miles of bus-only lanes throughout the city (City of Los Angeles 2022: 72–73). A behavior change campaign to encourage shared, sustainable mobility options was already implemented.

An evaluation of the measures is still hardly possible since most of the goals are only to be achieved in the medium and long terms. What is characteristic, however, is that a shift of trips from cars to buses and trains should be mainly achieved through a supply-oriented mobility pattern, such as an improved train network or special bus-only lanes. As mentioned, this policy was not successful for the last 30 years: The private car is still, like in the 1970s and 1980s, the favorite mode of transport and even the very recent transit ridership trends after enormous investments are revealing.

Another important chapter of the sustainability plan calls for increasing the percentage of electric or zero-emission vehicles in the city from 1.4% in 2018 to 25% by 2025, 80% by 2035, and 100% by 2050. Moreover, by 2030, all public transport should be 100% electrified. The following are some important milestones:

- Install 28,000 EV chargers by 2028.
- Develop a roadmap for Fossil Fuel Free Zone by 2021 and implement it by 2030.
- Electrify 10% of taxi fleet by 2022 and 100% by 2028.

In the "Annual Report of the Green Deal of 2021/2022", it is mentioned that Los Angeles surpassed its 10,000 commercial electric vehicle (EV) chargers goal two years ahead of schedule and is now home to over 16,800 commercial chargers (the most commercial EV chargers of any US city), including over 430 charging stations that have been mounted on street lights. Los Angeles further placed an order of 155 electric buses, which it claimed was the nation's largest single order (City of Los Angeles 2022: 26).

The new focus on electrification of cars seems to be much more promising for the car-dependent culture of Los Angeles than to convince residents, especially low-income residents, to rely on public transit. The number of registrations of all electric vehicles in the Los Angeles County has been rapidly rising in recent years. While the rate of growth in new registrations of BEVs was relatively modest up until 2020, it has been much more dynamic since then. In total, well over 200,000 pure electric vehicles will already be in stock in Los Angeles by the end of 2022 (see Figure 7.3).

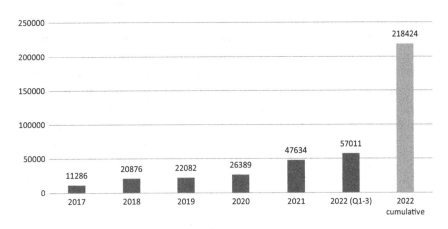

Figure 7.3:   Los Angeles county electric vehicles sales (BEV).

Source: California Energy Commission (2022); "2022 cumulative" includes all BEVs registered until September 2022.

While the change of fuel technology might have a positive impact on reducing CO_2 emissions, it will not solve the general mobility problems of Los Angeles, especially the most prevailing problem of congestion.

7.5 Autonomous driving as a solution to the mobility problems of Los Angeles?

Can autonomous driving solve the mobility problems in the city and improve the traffic situation in Los Angeles? Basically, the technological development of autonomous vehicles has made significant progress since 2010. Autonomous driving is differentiated according to different degrees or levels of autonomy.[12] With so-called Advanced Driver Assistance Systems (up to Level 2), the driver is supported while driving, for example, by active distance and lane keeping systems, but the driver still has full responsibility for the driving task and must monitor the traffic at any time. In contrast, from autonomous driving Levels 3 to 5, the vehicle assumes sole responsibility for the driving task in the defined driving situation. An example of a Level 3 system is the Traffic Jam Pilot, which, for example, takes over the driving task completely on freeways up to a defined speed. The driver can keep busy with other activities until the driving system calls on the driver to take over the driving task again. With the "Drive-Pilot", Mercedes-Benz was the first manufacturer to receive approval for such a system in Germany in 2022 and from 2023 also in the US (Mercedes-Benz 2023).

Fully autonomous driving systems at Levels 4 and 5 no longer require the driver to take over the vehicle. These include robotaxis or roboshuttles, which are already in commercial operation in cities such as Phoenix, Arizona (since 2016), San Francisco, and in China (e.g., Shanghai). In the US, Alphabet subsidiary Waymo, GM subsidiary Cruise, and Motional, a joint venture created by Aptiv and Hyundai, are considered pioneers (Bratzel/Tellermann 2022: 81–82). Cruise has been offering a commercial autonomous ride-hailing service in San Francisco since 2022, without a safety driver on board and is planning to increase the fleet of fewer than 100–5,000 cars (McFarland 2022). The future vision of many providers is that the travel costs of autonomous taxis and shuttles, without a human

---

[12]According to SAE, six levels of driving automation are defined: Level 0 (no driving automation) to Level 5 (full driving automation) (cf. SAE 2023).

driver, can be significantly reduced compared to conventional costs per mile of taxis (Litman 2023: 3).

Autonomous driving services will also be offered in Los Angeles soon. In fall 2022, Waymo announced that Los Angeles would be the next ride-hailing city after Phoenix and San Francisco to offer a round-the-clock service that provides more accessible and dependable mobility options to all residents of L.A. (Pearl 2022). In the press release, Waymo emphasized the city's economic importance for ride-hailing: "With approximately 13 million residents, the Los Angeles metropolitan area is one of the largest ride-hailing service areas in the world and the third largest in the US, with an estimated market opportunity of $2 billion in 2022" (Waymo 2022). In November 2022, Motional also announced that they, together with ride-hailing provider Lyft, would launch a robotaxi service in Los Angeles. It will be some time before the autonomous driving services can be rolled out widely. Permits still have to be obtained from the State Department of Motor Vehicles before the commercial services are gradually introduced (Hawkins 2022).

What influence will autonomous driving presumably have on traffic and mobility development in Los Angeles? From a political point of view, there are (again) high expectations for the new technologies of autonomous vehicles to solve the city's traffic problems. The mayor of Los Angeles puts it on record that "(b)y adding Waymo to our growing list of ways to get around, we're making good on our commitment to ease congestion on our streets, clean our air, and give people a better way to get where they need to go" (Waymo 2022). The statement impressively reflects the dominant supply-oriented mobility paradigm of those responsible. A change in "car culture" should be made possible by offering more transport alternatives — such as the new autonomous driving services.[13]

There are not yet many empirical studies on the effects of autonomous vehicles on traffic development in metropolises like Los Angeles.[14] Basically, for an initial assessment of the possible effects of autonomous vehicles on traffic development, a distinction must be made between

---

[13]"If we want to change the car culture in Los Angeles, we need to give Angelenos real alternatives to owning their own vehicle — including a world-class public transportation network, a range of active transportation options, and the convenience of mobility as a service across our City," said Mayor Eric Garcetti (Waymo 2022).

[14]Most studies model the possible effects and come to different conclusions, see, e.g., Fagnant/Kockelman (2015), Mouratidis *et al.* (2021).

commercial robotaxi or roboshuttle services and autonomous private vehicles. Commercial autonomous ride-hailing represents another transport alternative for Los Angeles and expands the spatial mobility options of the residents. In particular, people who cannot drive a car on their own can benefit from this, such as the elderly, children, students, or people with disabilities. Accordingly, new activities become possible or more attractive, which can lead to new traffic or a shift of trips from public transport to robotaxis.

Autonomous driving in private vehicles represents a significant improvement in comfort for drivers in addition to safety aspects. Instead of being occupied with the task of driving, the driver can devote himself to other activities such as working, e-commerce, or even relaxing while driving from a to b. Especially for the abovementioned exhausting long driving times due to the many traffic jams in Los Angeles, the attractiveness of one's own autonomous private vehicle increases significantly with autonomous driving. Thus, it is very likely that autonomous vehicles will increase the number of car journeys and increase vehicle kilometers driven.

For the metropolis of Los Angeles, a study that models the traffic development of autonomous vehicles concludes that the total number of trips increases by 9%, with a 13% growth in total car-like mode travel distance. After that, the transportation system would not benefit from the proliferation of private and shared autonomous vehicles. Accordingly, it would be crucial that autonomous driving is controlled by political interventions (He *et al.* 2022). These findings are also confirmed by other recent studies.[15]

Autonomous vehicles can further strengthen or solidify the existing long-distance mobility structures, especially in the Los Angeles region. The lack of stressful driving and the time gained through productive activities while driving make it even more attractive to live far away from work and other places of activity, which means that travel times and urban sprawl increases (Taiebat *et al.* 2018). As a result, new traffic is created

---

[15]"We find a clear synergistic effect between the impacts, leading to considerably higher levels of automobile use under automation; we predict car-km to increase by 56% for private ownership and 41% under shared ownership by 2050 when compared with business as usual. Total person-km rise by 39% and 32% respectively, while use of public transport, walking and cycling fall, indicating a move to a less sustainable future" (May *et al.* 2020: 127).

on the existing freeways, which will probably result in a downward spiral of worsening traffic congestion of freeways and again increasing travel times. Autonomous private vehicles will thus further strengthen the car-centric culture in Los Angeles.

A new technology, such as autonomous driving, will not solve the traffic problems of Los Angeles if the existing mobility and policy paradigm are retained. The only chance of improving the traffic situation is to replace the supply and capacity-oriented mobility policy model with demand-oriented and spatial-structural strategies. These include appropriate public policies and regulations, such as zoning, pricing, and urban design, to avoid the negative impacts of increased suburbanization and longer-distance travel.

7.6   Lessons learnt? Transport policy in Los Angeles 1990–2022

Los Angeles has not managed to solve the metropolis's traffic problems in the last 30 years. The predominantly supply-oriented mobility patterns were essentially retained with the change in strategy to rebuild a rail transport system for the region. In addition, the capacity of the freeway network continues to expand, albeit to a lesser extent. What lessons can be learned from the development of transport policy in Los Angeles over the past few decades?

Increasing the capacity of roads and highways to reduce congestion induces new traffic in the shortest possible time: spatial shift of car traffic from other routes, time shift of car trips, switch from other modes of transport to the car, new, previously not made car trips, and long-term emergence of distance-intensive spatial structures. Due to the new traffic, the newly created road capacity is overcompensated again in a very short time, which leads to further, usually even larger traffic jams. Los Angeles is a prime example of this pattern.

In a region geared toward the private car with long-distance mobility structures, the construction of an extensive public rail network does not lead to drivers switching to public transport. As long as the accessibility of activity locations (e.g., measured in travel time) is many times better on average by car than by public transport, a switch is unlikely. In Los Angeles, public transit ridership consists primarily of residents who have not yet been able to afford a car. With the socially desirable economic advancement, however, low-income groups acquire a private vehicle and henceforth turn away from public transport for their mobility needs.

The pattern of action of supply-oriented mobility policy is very stable over time and it is politically low in conflict. But it does not solve the region's transport problems. Through the significant expansion of transport infrastructure — first the rail infrastructure (from the 1890s), then the freeways (from the 1920s), and then the resurrection of the rail network again — Los Angeles created and partially worsened the traffic problems that it wanted to solve. Again and again, Los Angeles adopted problem-solving strategies based on the same paradigm: increasing the capacity of the network (transit, freeways). Supply-oriented policies are easier to implement politically, but they do not solve the region's congestion and environmental problems.

Demand-oriented policy strategies aimed at changing traffic behavior are more suitable for solving traffic problems, since they do not induce any new traffic. As a rule, however, they are politically very conflictual and difficult to implement. These include pricing strategies that lead to more efficient use of existing traffic capacity. User fees, like peak-hour road pricing, tolls, gas taxes, and parking fees, raise the cost of driving and offer positive and negative incentives, helping either to reduce the overall demand for driving or to manage the demand for driving during peak hours. User fees thus generate revenue while reducing the need for new capacity investments. These strategies deviate from the dominant supply-side mobility paradigm in Los Angeles and are the exception (Wachs *et al.* 2020: 5).[16] However, the Los Angeles County Metropolitan Transportation Authority approved a study in 2019 to assess the feasibility of reducing traffic by charging different prices at different times and places for driving on streets and roads in Los Angeles. In the same year, the Southern California Association of Governments (SCAG) released the results of its own long-term congestion pricing study which led SCAG to recommend a pilot project for West Los Angeles within a specific area they are calling the "GO Zone" (*ibid.*).

---

[16]Beyond a handful of toll roads in Orange County and High Occupancy Toll (HOT) lanes, Greater Los Angeles has rarely experimented with anything akin to congestion pricing. Local leaders since the 1920s have seen "congestion not as an excess of cars but as a scarcity of street space, to be remedied by the supply of street capacity" (Wachs *et al.* 2020: 5).

Patterns of action that address spatial and mobility structures generally have a high level of impact, since they address the underlying causes of traffic formation. However, they are not easy to alter and can only be changed in the long term. Los Angeles shows that the vicious circle of distance-intensive mobility structures is difficult to reverse and relatively inelastic. The long-term oriented choice for certain places of residence and company locations, as well as the existing construction and infrastructure, forms a normative force of the factual. Over time, these choices manifest themselves in lifestyles and in certain mobility and automotive cultures.

In the last 30 years, a polycentric spatial structure with a comparatively high regional population density has increasingly developed in Los Angeles due to population growth and the described mobility patterns. However, polycentricity reinforces auto-dependence since activity locations are distributed more evenly throughout the region as jobs and other destinations are spread over more locations. This makes it more difficult for public transit to serve the people and less attractive to use (Wachs *et al.* 2020). It is precisely the combination of multiple journeys to different destinations in widely separated places that increases the attractiveness of the car and also increases the likelihood of traffic jams. The age of autonomous driving will increase the attractiveness of the private car and increase distance-intensive mobility patterns. Roboshuttles will only contribute to solving the traffic problems in Los Angeles if they are combined with another level of thinking, a different political paradigm that sets itself apart from the dominant capacity-oriented pattern of action.

Chapter 8

Results and Outlook

The thesis formulated in the introduction that the often dichotomously juxtaposed means of transport, automobile and train, can develop similar mobility patterns was confirmed in this work. In a historical analysis, common features of the two motorized modes of transport could be identified in their effect on the mobility structure of the metropolis, on the mobility behavior of citizens, and on the mobility opportunities of social groups.

Although it is generally assumed that urban mass transport systems led to dense settlement patterns, the case of Los Angeles shows that trams and regional trains rather created the distance-intensive mobility structures that still dominate the metropolis today. When the automobile replaced the railways as the dominant mode of transport in the 1920s, the urban sprawl of the metropolis had long since been shaped. Nevertheless, the enormous flexibility of the new means of transport, the car, considerably strengthened and expanded the horizontal settlement structures. The sharp division of use — especially between living and working activities — also became clear during the *railway phase*. Above all, the regional railways of *Pacific Electric* made it possible to establish *bedroom communities* in the outskirts of the metropolis, which led to long daily commutes to jobs in the center. The sharp spatial separation of living and working and the suburban lifestyle strengthened with the advent of the car.

In parallel to the analysis of the interrelationships between transport modes and mobility patterns, the transport policy patterns in Los Angeles were described. It was found that supply-oriented transport policies since the 1920s have favored and supported distance-intensive mobility structures. The freeway network built in the 1950s and 1960s was unable to meet the mobility needs of its citizens, mainly because it induced a large proportion of new traffic itself. Therefore, policies that improve motorized transport have only a limited depth of impact in terms of solving mobility problems. This is truer the larger the area that is made more accessible.

Even the recent resurrection of a large rapid transit network in Los Angeles — which is the largest rail expansion program in the US — shows this narrowly modeled, supply-oriented pattern. It is very doubtful whether this huge investment in public transport will change the car-focused travel behavior or at least attract large numbers of passengers in the long term. The "transit blues" of declining passenger volumes of public transport between 2010 and 2019 and the increasing car ownership trends in Los Angeles are not very promising.

It is very likely that the car-oriented lifestyle in the metropolis will also be strengthened by the new technological trends of autonomous driving. Autonomous private vehicles make the long journeys in Los Angeles more attractive even and especially if the cars are increasingly stuck in traffic jams as the driver can devote himself or herself to other activities like working or relaxing. If robotaxis and roboshuttles are simply added as a new transport alternative, it will most likely induce new trips and shift passengers from public transport to the more convenient autonomous vehicles. As long as the capacity-oriented transport policy pattern remains, the traffic volumes will increase, and the distance-oriented metropolitan structure will be cemented. In contrast, greater depth of impact would be achieved by demand management measures, like road pricing, and especially mobility structure-oriented measures that aim at a densification and mixture of activity locations (living, working, consumption, leisure) and thus address the conditions under which traffic is generated.

The second part of the analysis explained elements of the transport policy development in Los Angeles. Again, the starting point was the relationship between transport modes, transport policies, and mobility patterns. It was shown that the decline of the railways in Los Angeles did not require a conspiracy of the automotive industry. Rather, the sprawling structures created by the railways themselves made their operation in the metropolis unprofitable when their monopoly on transport was broken by competition from other motorized modes of transport. For the car in particular, the low density of settlement was a conceivably good starting condition and one of the determining factors for its overwhelming success in Los Angeles.

L.A.'s supply-oriented transport policy also becomes more plausible against the background of mobility structures. The numerous municipalities scattered across the metropolitan area have a great (location-bound) interest in transport connections due to their spatial location. Their institutional autonomy gives them a high potential for conflict, which makes coordinated policies more difficult. Under these conditions, supply-oriented measures are more successful because they can distribute the (transport policy) benefits spatially and thus hardly trigger any conflicts.

Finally, sociocultural factors were used to analyze the subjective foundations of transport policies, mobility patterns, and modes of transport. The individualistic–privatist elements of California's political culture were used to illustrate both the great influence of economic interests on urban development and the reluctance of state actors to intervene in these processes in a planned manner. Sociocultural orientations also make more plausible the suburban lifestyle that developed very early and dominantly in Los Angeles, which still represents an essential element of the metropolis's mobility structure today. With the symbolic dimensions of the car and railways as transport modes, the irrational elements of mobility were pointed out in the end, which have had an influence on individual behavior and transport policy action that should not be underestimated. Since the 1920s, hopes of solving urban problems using the automobile grew. The bitter disappointment of these hopes was accompanied by a symbolic change. Now, the solution to the problems of the metropolis is projected onto the symbol of the railway, but without the willingness of citizens to accept individual restrictions on their transport-intensive lifestyle. Therefore, the contribution of the rapid transit network to solving the traffic problems will also be small.

What are the lessons of the *Los Angeles transport policy failure* for other cities and regions? First, the Southern California experience has led to widespread skepticism about policies that see the solution to mobility problems primarily in the expansion or improvement of motorized modes of transport. Supply-oriented policies have a traffic-generating effect, and the larger the spaces they open up, the more traffic they generate. The actually astonishing result of this work is that this applies not only to roads and motorways but in principle also to rail networks, even though their negative ecological consequences are smaller. This can only be seen, however, if the intrinsic development dynamics of transport policy processes are considered. The main problem of supply-oriented policies lies in their medium- and long-term effects on mobility behavior and spatial structures. The opening up of new areas or the improvement of existing transport links triggers systemic adaptation processes, as a result of which spatial and settlement structures change and mobility behavior levels off to a more distant level.

Second, this dynamic understanding of mobility enhances strategies that influence mobility structures. The case of Los Angeles is an extreme example of the devastating consequences of neglecting regional planning and land-use planning for mobility patterns. If decisions on the location of residential areas, industrial, and commercial settlements are left to the market, there is a considerable risk of externalization of the transport costs (e.g., for infrastructure measures) and ecological follow-up costs of increased traffic. This highlights large-scale investments such as industrial parks and leisure facilities "on greenfield sites", which will permanently change mobility structures over decades. If policymakers allow urban sprawl and dispersion of activities by failure to coordinate political action, the future scope for *sustainable mobility* is minimized. The depth of impact of all subsequent measures will be comparatively low.

Mobility structure-oriented policies aim to reduce the need for transport by reducing the spatial–functional differentiation of individual and social activities. In this sense, measures have a high depth of impact which allow for a densification and mixture of activities in cities, i.e., housing, work, consumption, and leisure time are brought together again in spatial terms so that they are all accessible by bicycle and on foot at best (city of short distances). In this sense, it is precisely those policy areas that are of major importance, which are hardly ever discussed in public and are also rather neglected politically: structural policy, regional planning, urban development, land-use planning, improvements to the living environment, etc.

Third, if these structural policies are to succeed, they must be complemented by demand-led policies, like pricing strategies that lead to more efficient use of existing traffic capacity. As the mobility experience in Los Angeles shows, even under the worst conceivable conditions, citizens' behavior can be influenced by positive and negative incentives ("push and pull"). From a transport policy point of view, it also seems worthwhile to strengthen the various forms of teleworking (telehome work, satellite offices) and to consider new working time models (e.g., four-day week).

Telework has brought the greatest change in mode of commuting in the last decades with the COVID-19 pandemic bringing about a boost in the number of teleworkers. Whether a high level of telecommuting will remain in the future is not clear yet. With the pandemic easing, more companies are calling workers back to the office again.

Finally, the mobility experience in Los Angeles illustrates the great influence of sociocultural factors on the mobility behavior of citizens and, not least, on the actions of transport policy actors. The values and attitudes toward mobility prevailing in the societies or social groups must be taken into account as essential framework conditions for policy. They are expressed, for example, in the respective lifestyles (e.g., residential and leisure patterns) or in symbolic dimensions of transport modes. Acknowledging the relevance of these comparatively stable subjective orientations does not necessarily mean that political action is meaningless. Rather, it follows that transport is a cross-cutting issue, and transport policy goes far beyond the provision of (transport) infrastructure. A *mobility policy* understood in this sense must be deeply interlinked with social, political–cultural, and economic–technical areas in order to achieve an ecologically sustainable effect, for example, with economic, technology, structural, and research policies as well as social and environmental policies. But social actors, such as companies, must also fulfill their social–ecological responsibility and make their contribution to *sustainable mobility*. Last but not least, it is up to each individual to keep the *freeways* clear *in his or her head* so that he or she does not spend his or her life in traffic jams. The wise turtle in Michael Ende's famous novel *Momo* already knew that "The slower you drive, the faster you get ahead".

Appendix

The operation of the electric railways in Los Angeles 1913–1933 (All following information is off: Fogelson 1967: 168–184.)

Year	Revenue passengers (in millions)	Operating revenue[a] (in millions)	Operating expenses[b] (in millions)	Total deductions[c] (in millions)	Net income (in millions)
		Los Angeles Railway			
1913	139.8	$7.0	$4.9	$1.5	$0.6
1914	135.0	6.8	4.7	1.6	0.5
1915	122.2	6.1	4.2	1.7	0.3
1916	117.3	5.9	4.0	1.6	0.3
1917	123.1	6.1	4.5	1.6	0.1
1918	130.4	6.6	5.5	1.5	[0.5]
		Pacific Electric Railway			
1913	68.7	$n.a.	$n.a.	$n.a.	$n.a.
1914	70.3	9.0	6.6	3.5	[0.6]
1915	64.7	n.a.	n.a.	n.a.	n.a.
1916	61.2	8.3	5.9	3.8	[1.0]
1917	65.0	9.0	6.3	4.0	[0.9]
1918	67.9	10.1	8.1	4.2	[1.7]

Notes: [a]Includes revenue from freight operations. [b]Includes depreciation.[c] Includes taxes. [ ] denotes deficit. n.a. denotes not available.
Source: Railroad Commission of the State of California, *Case No. 4002. Report on the Local Public Transportation Requirements of Los Angeles* (Los Angeles, 1935), chart facing p. 74, p. 82; Letter S. Ready, J. O. Marth, and Richard Sachte, *Joint Report on Street Railway Survey, City of Los Angeles* (Los Angeles, 1925), pp. 101, 171.

Year	Revenue passengers (in millions)	Operating revenue[a] (in millions)	Operating expenses[b] (in millions)	Total deductions[c] (in millions)	Net income (in millions)
			Los Angeles Railway		
1919	145.4	$7.3	$6.1	$1.6	$[0.3]
1920	179.2	9.0	7.2	1.6	0.2
1921	200.9	10.1	8.2	1.7	0.2
1922	219.0	11.1	7.6	2.2	1.4
1923	218.6	12.6	8.8	2.2	1.5
			Pacific Electric Railway		
1919	68.3	$11.3	$9.8	$4.4	$[2.8]
1920	84.5	15.3	12.0	4.6	[1.2]
1921	88.7	17.1	13.1	4.9	[0.8]
1922	88.1	18.3	13.8	5.1	[0.6]
1923	100.1	21.6	16.2	5.4	0.3

Notes: [a]Includes revenue from freight operations. [b]Includes depreciation. [c]Includes taxes. [ ] denotes deficit.
Source: Lester S. Ready, J. O. Marsh, and Richard Sachse, *Joint Report on Street Railway Survey, City of Los Angeles* (Los Angeles, 1925), pp. 101, 171; Railroad Commission of the State of California, *Case No. 4002. Report on the Local Public Transportation Requirements of Los Angeles* (Los Angeles, 1935), chart facing p. 74, p. 82.

Year	Revenue passengers (in millions)	Operating revenue[a] (in millions)	Operating expenses[b] (in millions)	Total deductions[c] (in millions)	Net income (in millions)
			Los Angeles Railway		
1924	255.6	$13.1	$9.9	$2.3	$1.1
1925	249.1	12.9	10.5	2.3	0.2
1926	250.8	13.0	10.9	2.3	—[d]
			Pacific Electric Railway		
1924	100.9	$20.7	$15.9	$4.7	$[0.6]
1925	94.8	19.5	16.0	4.1	—[d]
1926	92.8	19.1	16.4	4.2	[1.1]

Notes: [a]Includes revenue from freight operations. [b]Includes depreciation. [c]Includes taxes. [d]Less than $100,000. [ ] denotes deficit.
Source: Railroad Commission of the State of California, *Case No. 4002. Report on the Local Public Transportation Requirements of Los Angeles* (Los Angeles, 1935), chart facing p. 66, pp. 72, 74, 82.

Year	Revenue passengers (in millions)	Operating revenue[a] (in millions)	Operating expenses[b] (in millions)	Total deductions[c] (in millions)	Net income (in millions)
		Los Angeles Railway			
1927	254.5	$13.3	$11.1	$2.3	$0.1
1928	248.0	13.6	11.0	2.3	0.5
1929	223.7	14.9	11.2	2.4	1.5
		Pacific Electric Railway			
1927	91.6	$19.6	$16.4	$4.1	$[0.6]
1920	93.8	18.3	15.9	4.1	[1.2]
1929	97.0	18.4	15.5	4.0	[0.7]

Notes: [a]Includes revenue from freight operations. [b]Includes depreciation. [c]Includes taxes. [ ] denotes deficit.

Source: Railroad Commission of the State of California, *Case No. 4002. Report on the Local Public Transportation Requirements of Los Angeles* (Los Angeles, 1935), chart facing p. 66, pp. 72, 74, 82.

Year	Revenue passengers (in millions)	Operating revenue[a] (in millions)	Operating expenses[b] (in millions)	Total deductions[c] (in millions)	Net income (in millions)
		Los Angeles Railway			
1930	205.8	$13.7	$11.1	$2.4	$0.5
1931	182.5	12.2	10.7	2.2	[0.3]
1932	154.4	10.3	9.2	2.0	[0.5]
1933	140.5	9.5	8.3	1.9	[0.2]
		Pacific Electric Railway			
1930	90.5	$15.7	$14.0	$4.0	$[2.0]
1931	80.1	13.3	12.1	3.7	[2.3]
1932	67.1	10.5	10.0	3.4	[2.6]
1933	59.7	9.1	8.6	3.2	[2.6]

Notes: [a]Includes revenue from freight operations. [b]Includes depreciation. [c]Includes taxes. [ ] denotes deficit.

Source: Railroad Commission of the State of California, *Case No. 4002. Report on the Local Public Transportation Requirements of Los Angeles* (Los Angeles, 1935), chart facing p. 66, pp. 72, 74, 82.

Bibliography

Adler, S. (1986). The dynamics of transit innovation in Los Angeles. *Environment and Planning D: Society and Urban Space* 4: 321–335.

Adler, S. (1987). Why BART but no LART? The political economy of rail rapid transit planning in the Los Angeles and San Francisco metropolitan areas, 1945-57. *Planning Perspectives* 2: 149–174.

Adler, S. (1991). The transformation of the pacific electric railway. Bradford Snell, Roger Rabbit, and the politics of transportation in Los Angeles. *Urban Affairs Quarterly* 27: 51–86.

Allgemeiner Deutscher Automobilclub (ADAC). (1987). Mobility, Munich.

Altshuler, A. (1979). *The Urban Transportation System*. Cambridge (USA).

Altshuler, A., Anderson, M., Jones, D., Roos, D., and Womack, J. (1984). *The Future of the Automobile*. London, Sidney.

Armanski, G. (1986). *Die Lust und Last am Automobil* (The Desire and Burden of the Automobile). Stuttgart.

Automobile Club of Southern California. (1937). Traffic Survey, Los Angeles.

Automobile Club of Southern California. (1966). 1965 Los Angeles Metropolitan Travel Time Study, Western Section, Institute of Traffic Engineers, Los Angeles.

Bacharach, J. (1991). Das neue Verkehrskonzept von Los Angeles (The New Traffic Concept of Los Angeles). 2. Berliner Verkehrswerkstatt. Innenstadt-Konzept: Quote oder Management? (City Centre Concept: Quota or Management?). Organiser: Prof. Dr. Herwig Haase, Senator for Transport and Operations.

Baker, D. M. (1933). A Rapid Transit System for Los Angeles. Los Angeles.

Baldassare, M. (1986). *Trouble in Paradise. The Suburban Transformation in America*. New York.

Baldassare, M. (1991). Transportation in suburbia. Trends in attitudes, behavior and policy preferences in Orange county, California. *Transportation* 19: 207–222.

Banfield, E. C. (1974). *The Unheavenly City Revisited.* Boston.

Bannon, L. *et al.* (Eds.). (1982). *Information Technology. Impact on the Way of Life.* Dublin.

Bellah, R. N. *et al.* (1985). *Habits of the Heart. Individualism and Commitment in American Life.* New York.

Berger, B. M. (1969). *Working-Class Suburbs. A Study of Auto-Workers in Suburbia.* Berkeley.

Billerbeck, R. (1989). *Plebiszitäre Demokratie in der Praxis. Zum Beispiel Kalifornien* (Plebiscitary Democracy in Practice. For example California). Berlin.

Blanke, B. (Ed.). (1991). Stadt und Staat. Systematische, vergleichende und problemorientierte Analysen „dezentraler" Politik (City and State. Systematic, Comparative and Problem-Oriented Analyses of "Decentralised" Politics). *Politische Vierteljahresschrift*, Special Issue 22.

Bollens, S. A. (1992). State growth management. Intergovernmental frameworks and policy objectives. *American Journal of the American Planning Association* 58: 454–466.

Bottles, S. (1987). *Los Angeles and the Automobile. The Making of the Modem City.* Berkeley.

Bratzel, S. (1999a). Conditions of success in sustainable urban transport policy. Policy change in 'relatively successful' European cities. *Transport Reviews,* 2/1999, 19: S. 177–190.

Bratzel, S. (1999b). Erfolgsbedingungen umweltorientierter Verkehrspolitik in Städten. Analysen zum Policy-Wandel in den „relativen Erfolgsfällen" Amsterdam, Groningen, Zürich und Freiburg (i.Brg.) (Conditions of Success in Sustainable Urban Transport Policy. Policy Change in 'Relatively Successful' European Cities Amsterdam, Groningen, Zürich und Freiburg (i. Brg.)). Basel, Boston, Berlin: Birkhäuser Verlag 1999 (zugl. Dissertation).

Bratzel, S., and Tellermann, R. (2022). Mobility Services Report 2022. CAM Working Paper, Bergisch Gladbach, September 2022. Retrieved from: https://auto-institut.de/mobility-services-2/.

Brey, J. (2022). Why Denver and L.A. are backing away from highway expansions. *Governing.* Retrieved from: https://www.governing.com/now/how-new-climate-rule-stopped-highway-expansion-in-denver.

Brodsly, D. (1981). *L.A. Freeway. An Appreciative Essay.* Berkeley.

Brownell, B. (1972). A symbol of modernity: Attitudes toward the automobile in southern cities in the 1920s. *American Quarterly* 24: 20–44.

Bums, L. S., and Harman, A. J. (1968). The Complex Metropolis, Los Angeles.

Bureau of the Census (versch. Jahrgänge). Census of Population and Housing. Summary Population and Housing Characteristics, Washington.

California Energy Commission. (2022). New ZEV Sales in California. Retrieved from: https://www.energy.ca.gov/data-reports/energy-almanac/zero-emission-vehicle-and-infrastructure-statistics/new-zev-sales.

Cervero, R. (1986). Intrametropolitan trends in sunbelt and western cities: Transportation implications. *Transportation Research Record 1067*: *Social and Economic Factors in Transportation*. Washington.

Cervero, R. (1988). Land-use mixing and suburban mobility. *Transportation Quarterly* 42: 429–446.

Cervero, R. (1989a). *America's Suburban Centers. The Land Use-Transportation Link*. Boston.

Cervero, R. (1989b). Jobs-housing balancing and regional mobility. *Journal of the American Planning Association* 55(2): 136–150.

Cervero, R. (1991). Land uses and travel at suburban activity centers. *Transportation Quarterly* 45: 479–491.

Cerwenka, P. (1982). Personenverkehrsmobilität: Geschichte, Befunde und Ausblick (Passenger Transport Mobility: History, Findings and Outlook). *Report on the 9th International Symposium on Theory and Practice of Transport Science of the European Conference of Ministers of Transport*, Madrid, 2–4 November.

Christiansen, T. and Gordon, L. (1992). Assessment of the Small Employer Market. Commuter Transportation Services, Inc., Los Angeles.

City of Los Angeles. (2019). L.A.'s Green New Deal. Sustainable City Plan, Los Angles. Retrieved from: https://plan.lamayor.org/sites/default/files/pLAn_2019_final.pdf.

City of Los Angeles. (2022). L.A.'s Green New Deal. Sustainable City Plan, Annual Report 2021–2022, Los Angeles. Retrieved from: https://plan.lamayor.org/sites/default/files/GND_Annual_Report_2022.pdf.

Commuter Transportation Services. (1992). State of the Commute Report, Los Angeles.

Dear, M. (1992). Understanding the NIMBY syndrome. *American Journal of the American Planning Association* 58: 288–300.

Der Sachverständigenrat für Umweltfragen (The German Advisory Council on the Environment). (1994). Umweltgutachten 1994, Stuttgart.

Dewees, D. N. (1970). The decline of street railways. *Traffic Quarterly* 4: 563–581.

Didion, J. (1977). The diamond slow down. *Esquire* (August 1977): 35–37.

Eamst/Young. (1992). South Coast Air Quality Management District Regulation XV Cost Survey, Los Angeles.

Eco, U. (1977). *Zeichen. EInführung in einen Begriff und seine Geschichte* (Symbol. Introduction to a Term and Its History). Frankfurt/Main.

Edelman, M. (1990). *Politik als Ritual. Die symbolische Funktion staatlicher Instititutionen und politischen Handelns* (Politics as Ritual. The Symbolic Function of State Institutions and Political Action). Frankfurt.

Eyestone, R. (1971). *The Threads of Public Policy: A Study in Policy Leadership.* New York.

Fagnant, D. J. and Kockelman, K. (2015). Preparing a nation for autonomous vehicles: Opportunities, barriers and policy recommendations. *Transportation Research Part A: Policy and Practice* 77: 167–181. Retrieved from: https:// doi.org/10.1016/j.tra.2015.04.003.

Fine, H. (2021). Car-loving LA in midst of largest rail construction program in US. *Los Angeles Business Journal*, November 15. Retrieved from: https://labusinessjournal.com/infrastructure/car-loving-la-midst-largest-rail-construction-prog/.

Fischer, F. (1993). Bürger, Experten und Politik nach dem "Nimby"-Prinzip: Ein Plädoyer für die partizipatorische Policy-Analyse (Citizens, Experts and Politics according to the "Nimby" Principle: A Plea for Participatory Policy Analysis). In Héritier, A. (Ed.) *Policy Analysis. Critique and Reorientation*: pp. 451–470.

Flink, J. (1975). *The Car Culture.* Cambridge (USA).

Flink, J. (1988). *The Automobile Age.* Cambridge (USA).

Fogelson, R. (1967). *The Fragmented Metropolis. Los Angeles, 1850-1930.* Cambridge (USA).

Fogelson, R. M. (1971). *Violence as Protest. A Study of Riots and Ghettos.* New York.

Foster, M. (1976). The model-T, the hard sell, and Los Angeles' urban growth: The decentralization of Los Angeles during the 1920s. *Pacific Historical Review* 44: 459–484.

Foster, M. (1981). *From Streetcar to Superhighway: American City Planners and Urban Transportation.* Philadelphia.

General Motors. (1974). The Truth About 'American Ground Transport' — A Reply By General Motors, Submitted to the Subcommittee on Antitrust and Monopoly of the United States Senate, Washington.

Giddens, A. (1991). *The Consequences of Modernity.* Stanford.

Gordon, P. (1990). Modem cities and their economic role. *Prepared for the 25th Anniversary Conference.* Department of Urban and Regional Planning, Florida State University Tallahassee, Florida.

Gordon, P. and Richardson, H. W. (1992). Congestion, Changing Metropolitan Structure, and City Size in the U.S.: Some New Evidence, Los Angeles (unveröffentl. Manuskript).

Gordon, P., Richardson, H., and Jun, M.-J. (1991). The commuting paradox. Evidence from the top twenty. *American Journal of the American Planning Association* 57: 416–420.

Gottdiener, M. (Ed.). (1986). *Cities in Stress. A New Look at the Urban Crisis.* Beverly Hills, London.

Graehler, M., Mucci, R. A., and Erhardt, G. (2019). Understanding the recent transit ridership decline in major US cities: Service cuts or emerging modes? *Transportation Research Board 98th Annual Meeting,* Washington DC, United States. Retrieved from: https://trid.trb.org/view/1572517.

Guiliano, G. (1988). Testing the limits of TSM: The 1984 Los Angeles summer olympics. *Transportation* 15: 143–161.

Guiliano, G. and Small, K. (1990). Subcenters in the Los Angeles Region. The University of California Transportation Center, Berkeley, Working Paper, No. 39.

Guiliano, G. *et al.* (1991). Preliminary Evaluation of Regulation XV of the South Coast Air Quality Management District. The University of California Transportation Center, Working Paper No. 60.

Hägerstrand, T. (1987). Human Interaction and Spatial Mobility: Retrospect and Prospect. In Nijkamp, P., and Reickman, S. (Eds.) *Transportation and Planning in a Changing World.* Aldershot: pp. 11–28.

Hall, G. E. and Slater, C. M. (1992). *1992 County and City Extra, Annual Metro, City and County Data Book.* Lanham.

Hamer, A. M. (1976). *The Selling of Rapid Transit. A Critical Look at Urban Transportation Planning.* Lexington.

Hart, S. (1985). An Assessment of the Municipal Costs of Automobil Use, Pasadena (unveröfftl. Manuskript).

Hawkins, A. J. (2022). Motional and Lyft will launch a robotaxi service in Los Angeles. *The Verge,* November 17. Retrieved from: https://www.theverge.com/2022/11/17/23463403/motional-lyft-av-los-angeles.

He, B. Y., Jiang, Q., and Ma, J. (2022).Connected automated vehicle impacts in Southern California part-I: Travel behavior and demand analysis. https://www.sciencedirect.com/science/article/pii/S1361920922001572?via%3Dihub.

Heidenheimer, A. J., Heclo, H., and Adams, C. T. (1975). *Comparative Public Policy. The Politics of Social Choice in Europe and America.* New York.

Heimann, H. (1987). Massenmotorisierung — historischer Prozess (Mass Motorisation — Historical Process). In VDA (Ed.) Society and the Automobile. Opportunities, Risks and Requirements for Action. Frankfurt: pp. 19–32.

Heinze, W. G. and Kill, H. (1987). Chancen und Grenzen der neuen Informations- und Kommunikationstechniken. Zur Übertragung verkehrevolutorischer Erfahrungen auf die Telekommunikation (Chances and Limits of the New Information and Communication Technologies. On the Transfer of Traffic Evolutionary Experiences to Telecommunications). In Veröffentlichungen der Akademie für Raumforschung und Landesplanung. (Ed.) *Räumliche Wirkungen der Telematik,* Vol. 169: pp. 21–72.

Heydorn, H. H. (1980). *Probleme amerikanischer Neustädte* (Problems of American New Cities). Cologne.

Hilton, G. and Due, J. (1960). *The Electric Interurban Railways in America.* Stanford.

Hirsch, W. Z. (Ed.). (1971). *Viability and Prospects for Metropolitan Leadership.* New York.

Holzapfel, H. (Ed.). (1988a). *Ökologische Verkehrsplanung: Menschliche Mobilität, Strapenverkehr und Lebensqualität* (Ecological Transport Planning: Human Mobility, Road Traffic and Quality of Life). Frankfurt/Main.

Holzapfel, H. (1988b). *Die Wechselwirkung zwischen Raumerschließung und Raumzerstörung* (The Interaction between the Development and Destruction of Space). In Holzapfel, H. (Ed.): pp. 3–13.

Howell, D. W. and Cutler, C. D. (1990). *Freeway Fatal Accidents 1987.* Sacramento.

Hradil, S. (1992). Alte Begriffe und neue Strukturen. Die Milieu, Subkultur- und Lebensstilforschung der 80er Jahre (Old Concepts and New Structures. The Milieu, Subculture and Lifestyle Research of the 80s). In Hradil, S. (Ed.) *Zwischen Bewusstsein und Sein. Die Vermittlung "objektiver" Lebensbedingungen und "subjektiver" Lebensweisen* (Between Consciousness and Being. The Mediation of "Objective" Living Conditions and "Subjective" Lifestyles). Opladen: pp. 15–55.

Ingram, H. M., and Mann, D. E. (Eds.). (1980). *Why Policies Succeed or Fail.* Beverly Hills, London.

Ionescu, D. (2022). Rethinking Highway Expansions. Retrieved from: https://www.planetizen.com/news/2022/09/118947-rethinking-highway-expansions.

IQAIR. (2022). Air quality in Los Angeles. Retrieved from: https://www.iqair.com/usa/california/los-angeles.

Jänicke, M. (1990). Erfolgsbedingungen von Umweltpolitik im internationalen Vergleich (Conditions for the Success of Environmental Policy in International Comparison). *Zeitschrift für Umweltpolitik und Umweltrecht*: 13: 213–232.

Jones, D. (1985). *Urban Transit Policy. An Economic and Political History.* Englewood Cliffs (USA).

Jones, P. M. (1987). Mobility and the individual in western industrial society. In Nijkamp, P., and Reickman, S. (Eds.) *Transportation and Planning in a Changing World.* Aldershot: pp. 29–47.

Kamphausen, G. (1992). Ideengeschichte Ursprünge und Einflüsse (Origins and Influences in the History of Ideas). In Zöller, M. (Ed.) *Political Culture.* Adams, W. P. *et al. Country Report USA.* Bonn: pp. 259–280.

Kleinsteuber, H. J. (1977). Staatsintervention und Verkehrspolitik der USA: Die Interstate Commerce Commission. Ein Beitrag zur politischen Ökonomie der Vereinigten Staaten von Amerika (State Intervention and Transport Policy in the USA: The Interstate Commerce Commission. A Contribution to the Political Economy of the United States of America), Stuttgart.

Kleinsteuber, H. (1992). Die Verkehrspolitik (Transport Policy). In Spahn, B. (Ed.) Economic system and economic policy. Adams, W. P. *et al.*: *Country Report USA*. Bonn: pp. 714–718.

KPCC. (2018). What widening the 405 in Orange County means for drivers — and congestion. January 26. Retrieved from: https://www.kpcc.org/programs/take-two/2018/01/26/61404/expansion-of-the-405-in-orange-county-has-begun-he/?_gl=1*mehr78*_ga*ODUwOTY0NzEuMTY3Mjg0ODY00Q..*_ga_02V0FNLNZR*MTY3MzQzMjU4Ny40LjAuMTY3MzQzMjU4Ny42MC4wLjA).

LA Almanac. (2022). Motor Vehicle Registrations Los Angeles County. Retrieved from: http://www.laalmanac.com/transport/tr02.php.

Lee, D. (1987). *Highway Infrastructure Needs*. Cambridge.

Lewis, D., and Goldstein, L. (Eds.). (1980). *The Automobile and the American Culture*. The University of Michigan Press.

Liebert, R. J. (1976). *Disintegration and Political Action. The Changing Functions of City Governments in America*. New York.

Linder, W. *et al.* (1975). *Erzwungene Mobilität. Alternativen zur Raumordnung, Stadtentwicklung und Verkehrspolitik* (Forced Mobility. Alternatives to Regional Planning, Urban Development and Transport policy). Cologne.

Litman, T. (2023). Autonomous Vehicle Implementation Predictions.Implications for Transport Planning: Victoria Transport Policy Institute. Retrieved from: https://www.vtpi.org/avip.pdf.

Logan, J. R., and Molotch, H. L. (1987). *Urban Fortunes. The Political Economy of Place*. Los Angeles.

Los Angeles County Transportation Commission. (1992). 30-Year Integrated Transportation Plan. Los Angeles.

Lowi, T. J. (1964). Public policy, case-studies, and political theory. *World Politics* 16: 677–715.

Lowi, T. J. (1972). Four systems of policy, politics and choice. *Public Administration Review*: 32: 298–310.

Manville, M., Taylor, B. D., and Blumenberg, E. (2018). Falling Transit Ridership: California and Southern California. Prepared for the Southern California Association of Governments, January 2018, Los Angeles. Retrieved form: https://scag.ca.gov/sites/main/files/file-attachments/its_scag_transit_ridership.pdf.

Manville, M., Taylor, B. D., Blumenberg, E., and Schouten, A. (2023). Vehicle access and falling transit ridership: Evidence from Southern California. *Transportation* 50: 303–329. Retrieved from: https://link.springer.com/content/pdf/10.1007/s11116-021-10245-w.pdf?pdf=button.

Marchand, B. (1986). *The Emergence of Los Angeles. Population and Housing in the City of Dreams, 1940–1970.* London.

Massotti, L. H., and Hadden, J. K. (Eds.). (1973). *The Urbanization of the Suburbs.* Beverly Hills.

May, A., Shepherd, S., Pfaffenbichler, P., and Emberger, G. (2020). The potential impacts of automated cars on urban transport: An exploratory analysis. *Transport Policy* 98: 127–138. Retrieved from: 10.1016/j.tranpol.2020.05.007.

McCarty Carino, M. (2018). Nothing Can Fix LA Traffic, So Deal With It. *LAIST*, July 12. Retrieved from: https://laist.com/news/nothing-can-fix-la-traffic-so-deal-with-it.

McCone, J. A. (1966). Violence in the city — An end or a beginning? In Crump, S.: *Black Riots in Los Angeles. The Story of the Watts Tragedy.* Appendix: The Text of the McCone Commission Report, S. 125–154.

McDaniels, W. (1971). Re-evaluating Freeway Performance in Los Angeles, Los Angeles. MA-Thesis UCLA.

McElhiney, P. (1960). Evaluating freeway performance. *Traffic Quarterly* 14: 296–312.

McFarland, M. (2022). GM's Cruise wants to add 5,000 more robotaxis to American streets. This city warns it could backfire. *CNN*, September 30. Retrieved from: https://edition.cnn.com/2022/09/30/business/cruise-gm-san-francisco-self-driving/index.html.

McSpedion, E. (1989). Building a Light Rail Transit in Existing Rail Corridors — Panacea or Nightmare? The Los Angeles Experience. Transportation Research Board: Light Rail Transit. New System Success at Affordable Prices. Washington: pp. 426–441.

Melbeck, C. (1990). Die Machtstruktur deutscher und amerikanischer Städte in Abhängigkeit von institutionellen Rahmenbedingungen (The Power Structure of German and American Cities in Dependence on Institutional Frameworks). *APuZ* 25: 37–45.

Mercedes-Benz. (2023). Certification for SAE Level 3 system for U.S. market. Press Release, January 26. Retrieved from: https://group.mercedes-benz.com/innovation/product-innovation/autonomous-driving/drive-pilot-nevada.html.

Metro. (2020). Facts at a Glance. Data current as of November 2020. Retrieved from: https://web.archive.org/web/20210814041302/https://www.metro.net/news/facts-glance/.

Metro. (2021). Board-Report 2021. Retrieved from: https://boardagendas.metro.net/board-report/2021-0724/.

Metro. (2022). About Metro. Retrieved from: https://www.metro.net/about.

Metro. (2023). East San Fernando Valley Light Rail Transit Project. Retrieved from: https://www.metro.net/projects/east-sfv/.

Meyer, T. (1992). Die Inszenierung des Scheins (The Staging of the Appearance). Essay montage, Frankfurt/Main.

Meyer, J. R., and Gomez-Ibanez, J. A. (1981). *Autos, Transit, and Cities.* Cambridge (USA).

Meyer, J. R., Kain, J. F., and Wohl, M. (1966). *The Urban Transportation Problem.* Cambridge (USA).

Meyfahrt, R. (1988). Verringerte Mobilität — eine Katastrophe? (Reduced Mobility — A Catastrophe?). In Holzapfel, H. (Ed.): pp. 106–126.

Mokhtarian, P. L. (1988). An empirical evaluation of the travel impacts of tele-conferencing. *Transportation Research* 22A: 283–289.

Monheim, H., and Monheim-Dandorfer R. (1990). *Straßen für alle. Analysen und Konzepte zum Stadtverkehr der Zukunft* (Roads for Everyone. Analyses and Concepts for Urban Transport of the Future). Hamburg.

Motor Vehicle Manufacturer Association. (1977). Motor Vehicle Facts and Figures'77, Detroit.

Mouratidis, K., Peters, S., and Van Wee, B. (2021). Transportation technologies, sharing economy, and teleactivities: Implications for built environment and travel. *Transportation Research Part D: Transport and Environment* 92: 102716. Retrieved from: https://doi.org/10.1016/j.trd.2021.102716.

Muller, P. O. (1981). *Contemporary Suburban America.* Englewood Cliffs (USA).

Mumford, L. (1964). *The Highway and the City.* New York.

Nelson, H. J. (1983). *The Los Angeles Metropolis.* Los Angeles.

Nelson, H. J., and Clark, W. A. (1976). *Los Angeles. The Metropolitan Experience: Uniqueness, Generality, and the Goal of Good Life.* Los Angeles.

Nijkamp, P. (1987). Mobility as a Societal Value: Problems and Pardoxes. In Nijkamp, P., and Reickman, S. (Eds.) *Transportation and Planning in a Changing World.* Aldershot: pp. 73–90.

Nilles, J. M. *et al.* (1976). *The Telecommunication Transportation Trade-off.* New York.

Noortman, E. V. (1978). Travel as part of human activities: Towards an inte-gral behavioural approach. In Institute de Recherce des Transport (Ed.) *Mobility in Urban Life. Research Conference*, 28–30. September 1978 in Arc-et-Senans.

Nowak, H. (1987). Phänomen Mobilität (Phenomenon mobility). In VDA (Ed.) *Gesellschaft und Automobil. Chancen, Risiken und Handlungserfordernisse* (Society and the Automobile. Opportunities, Risks and Requirements for Action). Frankfurt: pp. 10–18.

OECD. (Ed.). (1988). *Cities and Transport. Athens, Gothenburg, Hong Kong, London, Los Angeles, Munich, New York, Osaka, Paris, Singapore.* Paris.

Owen, W. (1959). *Cities in the Motor Age.* New York.

Owen, W. (1966). *The Metropolitan Transportation Problem* (Revised Edition). Washington.

Owen, W. (1976). *Transportation for Cities. The Role of Federal Policy.* Washington.

Pahl, R. (1975). *Whose City?* Harmondsworth: Penguin.

Park, R. E. (1925). *The Urban Community as a Spatial Pattern and a Moral Order.* Publications of the American Sociological Association, Vol. 20: pp. 1–14.

Pearl, M. (2022). Waymo driverless rides are coming to Los Angeles. *Mashable,* October 19. Retrieved from: https://mashable.com/article/waymo-driverless-car-los-angeles.

Pegrum, D. F. (1968). *Transportation. Economics and Public Policy.* Homewood (USA).

Pendyala, R. M., Goulias, K. G., and Kitamura, R. (1992). Impact of telecommuting on spatial and temporal patterns of household travel. The University of California Transportation Center, Working Paper, No. 111.

Perloff, H. *et al.* (1973). *Prototype State of the Region: Report for Los Angeles County.* Los Angeles.

Pickrell, D. H. (1992). A desire named streetcar. Fantasy and fact in rail transit planning. *Journal of the American Planning Association* 58: 158–176.

Pikarski, A. E. (1987). *Commuting in America. A National Report on Commuting Patterns and Trends.* Westport (USA).

Prittwitz, V. (1990). *Das Katastrophenparadox. Elemente einer Theorie der Umweltpolitik* (The Disaster Paradox. Elements of a Theory of Environmental Policy). Opladen.

Prittwitz, V. V., and Wolf, K. D. (1993). Die Politik der globalen Güter (The Politics of Global Goods). In Prittwitz, V. V. (Ed.) *Umweltpolitik als Modernisierungsprozess. Politikwissenschaftliche Umweltforschung und Lehre* (Environmental Policy as a Process of Modernisation. Environmental Research and Teaching in Political Science). Opladen: pp. 193–218.

Prittwitz, V. V., Bratzel, S., Wegrich, K., and Bernhard R. (1992). Symbolische Umweltpolitik. Eine Sachstands- und Literaturstudie unter besonderer Berücksichtigung des Klimaschutzes, der Kernenergie und Abfallpolitik (Symbolic Environmental Policy. A Study of the State of Affairs and Literature with Special Reference to Climate Protection, Nuclear Energy and Waste Policy). In Forschungszentrum Jülich (Ed.) *Arbeiten zur Risiko-Kommunikation.* Jülich, Vol. 34.

Pross, H. (1974). *Politische Symbolik. Theorie und Praxis der öffentlichen Kommunikation* (Political Symbolism. Theory and Practice of Public Communication). Stuttgart.

Richmond, J. (1991). *Transport of Delight — The Mythical Conception of Rail Transit in Los Angeles.* Library of the SCRTD, Los Angeles.

Rubin, T. A., and Moore, J. E. (2019). Metro's 28 by 2028 Plan: A Critical Review Metro's Transit Ridership Is Declining, March: Reason Foundation.

Retrieved from: https://reason.org/wp-content/uploads/la-metro-transit-ridership-is-declining.pdf.

Sachs, H. (1984). *Die Liebe zum Auto* (The Love of Cars). Reinbeck.

SAE. (2023). SAE Levels of Driving Automation. Refined for Clarity and International Audience. Retrieved from: https://www.sae.org/blog/sae-j3016-update.

Sampath, S., Saxena, S., and Mokhtarian, P. L. (1991). The Effectiveness of Telecommuting as a Transportation Control Measure. The University of California Transportation Center, Working Paper, No. 78.

SCAG. (1989a). Regional Growth Management Plan, Los Angeles.

SCAG. (1989b). Regional Mobility Plan, Los Angeles.

SCAG. (2020). Regional Guide 2020. Retrieved from: https://scag.ca.gov/sites/main/files/file-attachments/scag-2020-regional-guide.pdf.

SCAG. (2021). Regional Guide 2021. Retrieved from: https://scag.ca.gov/sites/main/files/file-attachments/scag-2021-regional-guide.pdf.

SCAQMD. (1991a). 1991 Air Quality Management Plan. South Coast Air Basin, Los Angeles.

SCAQMD. (1991b). Summary of Air Quality in California's South Coast and Southeast Desert Air Basins, Los Angeles.

Schnalzer, R. (2021). Traffic is terrible again. Here's how to get it closer to spring 2020 levels. *LA Times*, July 22. Retrieved from: https://www.latimes.com/business/story/2021-07-22/los-angeles-traffic-congestion-commute-pandemic.

Seelye, H. (1984). Few will pay for new roads. *California Journal*, 428–431.

Sharp, S. (2022). Eastside rail extension likely to start with first phase to Montebello. *Urbanize*, November 14. Retrieved from: https://la.urbanize.city/post/eastside-rail-extension-likely-start-first-phase-montebello.

Shevky, E. and Williams, M. (1949). *The Social Areas of Los Angeles. Analysis and Typology*. Berkeley.

Sloterdijk, P. (1992). Die Gesellschaft der Kentauren. Philosophische Bemerkungen zur Automobilität (The Society of Centaurs. Philosophical Remarks on Automobility). *FAZ-Magazin*, from 24.4.: 28–38.

Snell, B. (1973). American Ground Transport, Appendix to US, Congress, Senate, Committee on the Judiciary: The Industrial Reorganization Act: Hearings before the Subcommittee of the Senate Committee on the Judiciary on S. 1167, Part 4A, 93rd, Congress, 2nd Session, 1974.

Snell, B. (1974). Statement of Bradford Snell before the United States Senate Subcommittee on Antitrust and Monopoly, Presented on Hearings on the Ground Transportation Industries, in Connection with S. 1167, Washington.

Soja, E., Morales, R., and Wolff, G. (1983). Urban restructuring: An analysis of social and spatial change in Los Angeles. *Economic Geography* 59: 195–229.

Sorensen, P., Wachs, M. *et al.* (2008). *Moving Los Angeles. Short-Term Policy Options for Improving Transportation.* RAND Corporation, Santa Monica.

Sorokin, P. A. (1959). *Social and Cultural Mobility.* Glencoe (USA) (1st edn. 1927).

Sotero, D. (2013). First phase of Metro Red Line celebrates 20-year anniversary, Metro. *The Source.* Retrieved from: https://thesource.metro.net/2013/01/29/metro-rail-at-20-2/.

Southern California Commuter Rail. (1991). 1991 Regional System Plan, Los Angeles.

St. Clair, D. (1981). The motorization and decline of urban public transit, 1935–1950. *Journal of Economic History* 16: 579–600.

St. Clair, D. (1986). *The Motorization of American Cities.* New York.

Statista. (2022). Annual ridership of Los Angeles' transit authority, by mode 2015–2021. Retrieved from: https://www.statista.com/statistics/1297553/lacmta-network-total-annual-ridership-by-mode/.

Strauss, A. (1976). *Images of the American City.* New Brunswick (USA).

Taiebat, M., Brown, A., Safford, H., Qu, S., and Xu, M. (2018). A review on energy, environmental, and sustainability implications of connected and automated vehicles. *Environmental Science & Technology* 52 (20): 11449–11465. Retrieved from: 10.1021/acs.est.8b00127.

Thomson, M. J. (1978). Grundlagen der Verkehrspolitik, mit einem Vademecum zur Verkehrswirtschaft und Verkehrspolitik (Fundamentals of Transport Policy, with a Vademecum on Transport Economics and Transport Policy) by G. Wolfgang Heinze, Bern, Stuttgart.

TomTom. (2022). TomTom Traffic Index Ranking 2021. Retrieved from: https://www.tomtom.com/en_gb/traffic-index/los-angeles-traffic.

Transportation Engineering Board. (1939). A Transit Program for the Los Angeles Metropolitan Area. Presenting Recommendations for Development of Facilities for Private and Mass Transit and a Plan for Coordination of Mass Transit Operations, Los Angeles.

U.S. Census Bureau. (2020). Measuring America's People, Places, and Economy. Retrieved from: https://www.census.gov/.

Ullrich, O. (1988). *Industrielle Lebensweise udn das Automobil* (Industrial Lifestyle and the Automobile). In Holzapfel, H. (Ed.): pp. 14–36.

Uranga, R. (2022). Rail line in southeast L.A. County approved as leaders seek to speed up construction. *Los Angeles Times,* January 28, 2022. Retrieved from: https://www.latimes.com/california/story/2022-01-28/southeast-los-angeles-metro-rail-project-approved.

Verkoren, O. and Weesep, J. V. (Eds.). (1987). *Mobility and Uiban Change.* Amsterdam.

Voigt, R. (1989). Mythen, Rituale und Symbole in der Politik (Myths, Rituals and Symbols in Politics). In Voigt, R. (Ed.) *Politik der Symbole, Symbole der Politik* (Politics of Symbols, Symbols of Politics). Opladen.

Vorländer, H. (1992). Empirische Aspekte der politischen Kultur (Empirical Aspects of Political Culture). In Zöller, M. (Ed.) *Politische Kultur.* Adams, W. P. *et al.* Country Report USA. Bonn: pp. 303–327.

Wachs, M. (1976). Consumer attitudes towards transit service: An interpretative review. *Journal of the American Institute of Planners* 42: 96–104.

Wachs, M. (1984). Autos, transit, and the sprawl of Los Angeles: The 1920s. *Journal of the American Planning Association* 50: 297–310.

Wachs, M. (1990). Regulation traffic by controlling land use: The southern California experience. *Transportation* 16: 241–256.

Wachs, M. and Guiliano, G. (1992a). Employee transportation coordinators: A new profession in southern California. *Transportation Quaterly* 46: 411–427.

Wachs, M. and Guiliano, G. (1992b). Transportation policy options for southern California. Paper Prepared for Presentation at the Symposium: Policy Options for Southern California, November 19, 1992 (Unpublished manuscript).

Wachs, M. *et al.* (2020). A Century of Fighting Traffic Congestion in Los Angeles 1920-2020, September 2020, Los Angeles: UCLA. Retrieved from: https://luskincenter.history.ucla.edu/wp-content/uploads/sites/66/2020/10/A-Century-of-Fighting-Traffic-Congestion-in-LA.pdf.

Warner, S. B. Jr. (1962). *Streetcar Suburbs. The Process of Growth in Boston, 1870-1910.* Cambridge (USA).

Warner, S. B. Jr. (1972). *The Urban Wilderness. A History of the American City.* New York.

Wasserman, J. L. and Taylor, B. D. (2022). Transit blues in the golden state: Regional transit ridership trends in California. *Journal of Public Transportation*, 24. Retrieved from: https://doi.org/10.1016/j.jpubtr.2022.100030.

Waymo. (2022). Next Stop for Waymo One: Los Angeles, Waypoint. The official *Waymo* blog, October 19. Retrieved from: https://blog.waymo.com/2022/10/next-stop-for-waymo-one-los-angeles.html.

Weekly, R. L. (1990). *The Los Angeles County Transportation Commission and Regulation XV.* Los Angeles.

Weiner, E. (1987). *Urban Transportation Planning in the United States. An Historical Overview.* New York.

Weizäcker, E. U. (1990). *Erdpolitik. Ökologische Realpolitik an der Schwelle zum Jahrhundert der Umwelt* (Earth Politics, Ecological Realpolitik on the Threshold of the Century of the Environment). Darmstadt.

Whitt, A. J. and Yago, G. (1985). Corporate strategies and the decline of transit in U.S. cities. *Urban Affairs Quarterly* 21: 37–65.

Wilson, J. Q. (1967). A guide to reagan country. The political culture of southern California. *Commentary* 1967: 37–45.

Wolf, W. (1987). *Eisenbahn und Autowahn* (Railway and Car Mania). Hamburg.

Woodhull, J. (1991). Calmer, Not Faster: A New Direction for the Streets of L.A. *Prepared for the 70th Annual Meeting Transportation Research Board*, January 13–17, 1991, Washington.

Woodhull, J. (1992). How Alternative Forms of Development Can Reduce Traffic Congestion. In Walter, B. *et al. Sustainable Cities. Concepts and Strategies for Eco-city Development.* Los Angeles, S. 168–180.

Zöller, M. (1992). Politische Kultur und politische Soziologie (Political Culture and Political Sociology). In Zöller, M. (Ed.) *Politische Kultur.* Adams, W. P. *et al.: Länderbericht USA.* Bonn: pp. 281–302.

Index

Printed in the United States
by Baker & Taylor Publisher Services